3rd Congress on Materials Science and Engineering

(CNCIM-Mexico 2012)

IOP Conference Series: Materials Science and Engineering Volume 45

Yucatan, Mexico
27 February – 2 March 2012

Editors:

R. de Coss
G. Murrieta-Hernandez
A. Aguayo-Gonzalez
E. Rubio-Rosas
E. Chigo-Anota
E. Vigueras-Santiago

ISBN: 978-1-63439-856-5
ISSN: 1757-8981

Printed from e-media with permission by:

Curran Associates, Inc.
57 Morehouse Lane
Red Hook, NY 12571

Some format issues inherent in the e-media version may also appear in this print version.

Printed by Curran Associates, Inc. (2015)

For permission requests, please contact the Institute of Physics
at the address below.

Institute of Physics
Dirac House, Temple Back
Bristol BS1 6BE UK

Phone: 44 1 17 929 7481
Fax: 44 1 17 920 0979

techtracking@iop.org

Additional copies of this publication are available from:

Curran Associates, Inc.
57 Morehouse Lane
Red Hook, NY 12571 USA
Phone: 845-758-0400
Fax: 845-758-2634
Email: curran@proceedings.com
Web: www.proceedings.com

TABLE OF CONTENTS

SYNTHESIS AND PREPARATION OF MATERIALS: ORGANIC AND INORGANIC

CHARACTERIZATION OF MATERIALS: NOVEL METHODS AND TECHNIQUES

APPLICATIONS OF MATERIALS: ENVIRONMENT, MEDICINE, PHARMACY, TECHNOLOGY, FOOD AND RENEWABLE ENERGY

NEW MATERIALS: COMPOSITES, NANOSTRUCTURES, AND FROM NATURAL SOURCES

THEORY: NEW METHODS AND COMPUTER SIMULATIONS

PREFACE

The Third Congress on Material Science and Engineering (CNCIM-México 2012), which took place in Mérida, México, from 27 February to 2 March 2012 was organized by three research groups (cuerpos académicos) from the Universidad Autónoma de Yucatán: Ingeniería Física (UADY-CA-27), Modelado y Simulación Computacional de Sistemas Físicos (UADY-CA-101) and Química Fundamental y Aplicada (UADY-CA-32), in collaboration with the Centro de Investigación y de Estudios Avanzados (Cinvestav-Mérida).

The First Congress in Material Science and Engineering (CNCIM-2010), was organized in Puebla, México in February 2010. This was followed by CNCIM-2011 held in Toluca, México in February 2011.

The CNCIM-México 2012 Conference consisted of plenary talks (8), invited talks (10), oral contributions (54) and poster presentations (70). The topics of the Conference were:

- Synthesis and Preparation of Materials: Organic and Inorganic
- Characterization of Materials: Novel Methods and Techniques
- Applications of Materials: Environment, Medicine, Pharmacy, Technology, Food and Renewable Energy
- New Materials: Composites, Nanostructures, and from Natural Sources
- Theory: New Methods and Computer Simulations

We want to thank the Organizing Committee, the Institutions and Sponsors supporting the Conference, and everyone who contributed to the organization of this meeting, for their invaluable efforts in order to guarantee the complete success of this conference.

Editors

Romeo de Coss
Centro de Investigación y de Estudios Avanzados del I.P.N. (Cinvestav-Mérida)
A.P. 73 Cordemex 97310, Mérida, Yucatán, México
decoss@mda.cinvestav.mx

Gabriel Murrieta-Hernández
Universidad Autónoma de Yucatán
Calle 60 No. 491-A, Centro Histórico, C.P. 97000, Mérida, Yucatán, México
murrieta@uady.mx

Aarón Aguayo-González
Universidad Autónoma de Yucatán
Calle 60 No. 491-A, Centro Histórico, C.P. 97000, Mérida, Yucatán, México
aguayo@uady.mx

Efraín Rubio-Rosas
Benemérita Universidad Autónoma de Puebla

Ciudad Universitaria, Col. San Manuel, C.P. 72570, Puebla, Puebla, México
efrain.rubio@cuv.buap.mx

Ernesto Chigo-Anota
Benemérita Universidad Autónoma de Puebla
Ciudad Universitaria, Col. San Manuel, C.P. 72570, Puebla, Puebla, México
ernesto.chigo@correo.buap.mx

Enrique Vigueras-Santiago
Universidad Autónoma del Estado de México
Instituto Literario No. 100, Col. Centro 50000, Toluca, Edo. de México, México
vigueras@uaemex.mx

Session Chairs
Gabriel Canto Santana, Universidad Autónoma de Campeche.
Enrique Vigueras Santiago, Universidad Autónoma del Estado de México.
César Cab, Universidad Autónoma de Yucatán.
Alejandro ávila Ortega, Universidad Autónoma de Yucatán.
Jesús Barrón Zambrano, Universidad Autónoma de Yucatán.
Maritza de Coss, Universidad Autónoma de Yucatán.
Jorge A. Tapia González, Universidad Autónoma de Yucatán.
David Muñoz Rodríguez, Universidad Autónoma de Yucatán.
Mario Pérez Cortes, Universidad Autónoma de Yucatán.
Jesús García Serrano, Universidad Autónoma del Estado de Hidalgo.
Rubén Arturo Medina Esquivel, Universidad Autónoma de Yucatán.
César R. Acosta, Universidad Autónoma de Yucatán.

Organizing Committee
Aarón Aguayo González, Universidad Autónoma de Yucatán.
Gabriel Murrieta Hernández, Universidad Autónoma de Yucatán.
Alejandro Tapia González, Universidad Autónoma de Yucatán.
Cristian Carrera Figueiras, Universidad Autónoma de Yucatán.
Heriberto Hernández Cocoletzi, Benemérita Universidad Autónoma de Puebla.
Ernesto Chigo Anota, Benemérita Universidad Autónoma de Puebla.
Efraín Rubio Rosas, Benemérita Universidad Autónoma de Puebla.
Enrique Vigueras Santiago, Universidad Autónoma del Estado de México.
Romeo de Coss, Centro de Investigación y de Estudios Avanzados (Cinvestav-Mérida).

Organizers:

Sponsors:

IOP Conf. Series: Materials Science and Engineering **45** (2013) 012001 doi:10.1088/1757-899X/45/1/012001

Dinuclear transition metal complexes in carbon nanostructured materials synthesis

J I Ayuso [1,2,3], E Hernández [1] and E Delgado [1]

[1] Department of Inorganic Chemistry, Universidad Autónoma de Madrid. Cantoblanco 28049 Madrid, Spain

[2] Department of Inorganic Chemistry, Universidad de Alcalá. Campus Universitario Alcalá de Henares 28871 Madrid, Spain

[3] Present address: School of Chemistry and Organic Materials Innovation Centre, University of Manchester. Oxford Road M13 9PL Manchester, United Kingdom

E-mail: josue.ayuso@edu.uah.es

Abstract. Carbon nanomaterials (CNMs) were prepared with two similar techniques using organometallic complexes as catalysts precursors. Chemical vapour deposition (CVD) and pyrolysis with chlorine gas approaches were employed in order to explore the effect of dinuclear transition metal compounds [$Fe_2(CO)_6(\mu\text{-}S_2C_6H_2X_2)$, (X=OH, Cl)] in synthesis of CNMs. Our to-date results have shown these complexes generate different carbonaceous materials when they are used in bulk, it was also observed that their performances in synthesis differ even though these compounds are analogous. With X=OH complex used in CVD process, metal nanoparticles of ca. 20-50 nm in size and embedded in carbon matrix were obtained. X=Cl complex has been used in pyrolysis experiments and showed an entire volatilisation or no reaction, depending on selected temperature. Furthermore, obtaining of a new tetranuclear iron cluster is presented in this work.

1. Introduction

Nanochemistry devotes with synthetic aspects of the chemistry concerning the production and characterization of materials at the nanoscale [1]. Among the wide spectrum of those, carbon nanomaterials (CNMs) which include carbon nanotubes (CNTs), nanofibres, hollow and filled spheres, etc [2] are generally produced from catalytic decomposition of hydrocarbon gases over selected metal nanoparticles. In this sense, since the growing use of organometallic complexes in synthesis of CNTs as precursor catalysts [3], many studies have recently been done. Most of the study cases present the utilisation of simple organometallic compounds such as ferrocene or $Fe(CO)_5$, however study of ligand effects on CNM synthesis has not been emphasized [2].

Current literature points out that chemical vapour deposition (CVD) technique is the most common used approach to make CNMs due to its capacity to generate various types of nanostructures in a controlled manner and at high yield [4, 5]. A similar approach which involves the chlorination of organometallic compounds has also been explored as an alternative to synthesize CNMs [6].

Organometallic complexes generally undergo decomposition or pyrolysis reactions which make them suitable for the above mentioned methods. Also, they offer in a single molecule carbon source for the creation of carbon structures by means of the ligands, metal atoms for nanoparticle metal catalyst, and other atomic species or functional groups which could play different roles according to their nature.

In this sense, we explored the performance of dinuclear organometallic complexes and the effect of their ligands in synthesis of CNMs by using two different techniques.

2. Experimental Procedure

Organometallic complexes [$Fe_2(CO)_6(\mu\text{-}S_2C_6H_2X_2)$, (X=OH, Cl)] were synthesised using Schlenk techniques under Ar atmosphere. Compounds **1**, $Fe_2(CO)_6[\mu\text{-}S_2C_6H_2(OH)_2]$ and **2**, $Fe_2(CO)_6(\mu\text{-}S_2C_6H_2Cl_2)$, were prepared based on literature procedures [7, 8], respectively. However, while purifying the latter, in a subsequent elution with a mixture of n-hexane/CH_2Cl_2 (10:1) a new compound was unexpectedly isolated, **3**, in a 1:10 ratio respect to compound **2**, as a dark brown solid. **IR** (n-hexane) υ_{CO}/cm^{-1}: 2077(s), 2054(vs), 2022(m), 2015(s), 1977(s); $\upsilon_{C=C,\ aromatic}/cm^{-1}$: 1654(m). **^{1}H-RMN** ($CDCl_3$, 300 MHz, 22°C) δ_{ppm}: 7.10 (s, 4H, C-H). **MS (FAB$^+$)** *m/z*: 781.5-669.5 [M^+ – nCO; n=5-9]. **Elemental Analysis**: Calculated to $C_{20}H_4S_4Cl_4Fe_4O_8$ (Found): C, 26.53 (26.65); H, 0.64 (0.76); S, 13.50 (13.96).

CVD experiments consistent of two steps were carried out using a horizontal CVD reactor: Catalyst precursor preparation and CNMs growth. First, thermal degradation of compound **1** (0.02 mmol) was performed by placing it in bulk on a quartz vial and heated in air at 1073.15 K for 15 min in the CVD oven. Second, thermal CVD synthesis of CNMs was performed under a constant flow pressure by flowing a mixture of H_2 (1500 sccm) and of Ar (200 sccm) currents at 1073.15 K for 20 min to ensure reduction of iron. Subsequently, conditions were readjusted by keeping constant the H_2 and Ar flows and adding to this mixture a flow of ethylene (2 sccm) for 5 min. Finally, the H_2 flow was reduced 100 sccm every 5 min until a final flow of 200 sccm was reached while maintaining the Ar and ethylene flows. Then reactor and products were cooled down to room temperature before microscopy analysis.

Pyrolysis experiments consisted in chlorination of compound **2** (0.02 mmol), placed in a 6 cm long quartz boat, by treatment with a flow of ultrapure Cl_2 gas in a tubular furnace. The samples were heated at two different temperatures and heating ramps (1173.15 K, 50 K/min and 473.15 K, 10 K/min, respectively) during 30 min under a constant flow pressure by flowing a mixture of Cl_2 (25 cm^3/min) and Ar (50 cm^3/min). NaOH dissolution was connected at the end of the furnace to eliminate the unreacted Cl_2 and the generated volatile metal chlorides byproducts.

The general morphology of the obtained samples was observed with a Philips CM200 FEG (operated at 120 kV) transmission electronic microscope equipped with an EDAX DX4 spectrometer for XEDS by preparation of ultrasonic dispersions of samples in acetone. One drop of this suspension was deposited over a copper grid covered with a holey carbon film. Compounds **1**, **2** and **3** were analysed using Perkin Elmer Spectrum BX FT-IR, Bruker AMX-300 NMR, VG Autospec (Waters) MS using NBA as matrix in FAB$^+$ experiments, and LECO CHNS-932 Elemental Analysis.

3. Results and Discussion

Compound **1** used as precursor of CNMs by CVD process led to two types of carbon aggregates: embedding carbon and irregular unshaped carbonaceous material. As shown in figure 1, the sample is composed by metal nanoparticles of spherical shape and size of ca. 20-50 nm, embedded in carbonaceous matrix (a) and most of the irregular carbonaceous material is empty. XEDS analysis of metal particle marked with the arrow showed that it is composed of Fe, although traces of C and O is detected too, and is correspondent to the axis zone [001] of α-Fe phase (b). A magnification of the area marked with the circle is shown in (c), and we can see that spherical iron nanoparticles are encapsulated by an amorphous carbon wall and which seem to be composed of some carbonaceous layers. Moreover, the same experiment carried out in absence of ethylene only gave a rusty powder with no carbon content which would indicate that organic part of **1** is not a significant carbon source.

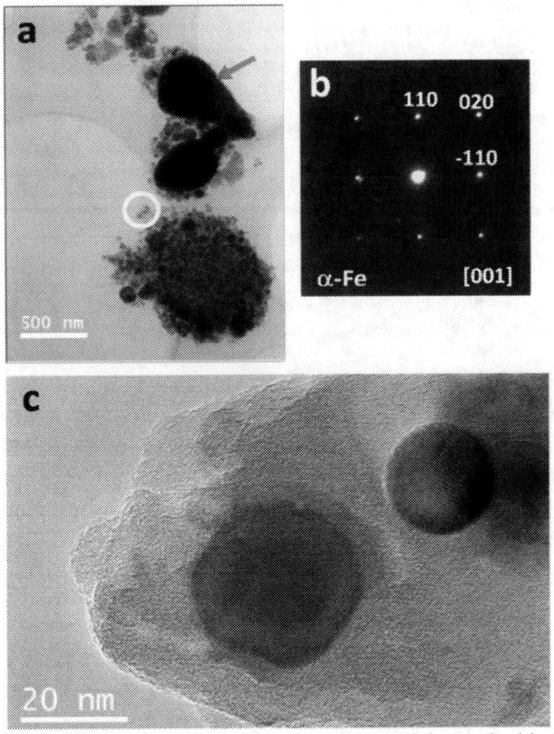

Figure 1. TEM micrograph of metal nanoparticles embedded in carbonaceous matrix (a). Electron diffraction pattern of the selected area (arrow) composed of α-Fe (b). Magnification of encircled area (c).

These results are in contrast with our previous studies [9] employing **1** in CVD process by using the same experimental conditions but with **1** previously deposited in silica substrates instead of in bulk. When deposited, nanoparticles generated by **1** are homogeneously distributed in the substrate surface which promotes a uniform catalytic activity in the CNTs growth. According to our results reported here they could be explained by the smaller number of active sites when precursor is treated in bulk, which would lead to the formation of nanoparticles with an irregular packing and subsequently, they could not be enough to achieve the growth of carbon nanotubes. In this case, if we consider the carbon from the ligands as an impurity, then it could be explained the presence of metal nanoparticles surrounded by carbon. Although it seems that the presence of oxygen could take an unusual role in its reaction performance by poisoning the iron catalyst [10] the presence of sulphur near the iron atoms could play a more significant role in the formation of carbon nanomaterials [11].

Attempts to obtain CNMs by using **2** through pyrolysis experiments failed under the used experimental conditions. When performed with the common temperature used in this type of experiments, 1173.15 K, we observed an entire volatilisation of precursors and no type of carbon soot. It is possible that in addition to the expected $FeCl_3$ byproduct, also another kind of chlorination process would be consuming the carbonyl and aromatic moieties, may be in the form of CCl_4 [12]. In contrast, at a temperature of 473.15 K there was not reaction, which could be attributed to the molecular weight of precursor as compared when organometallic complexes are mononuclear [6]. Both results suggest that the formation of CNMs would be achieved in temperatures between the tried ones, more likely near the 1173.15 K. New experiments are currently in process.

Characterisation of the unexpectedly obtained compound **3** indicated that is a tetranuclear iron cluster likely with a tetrahedral geometry containing nine terminal carbonyl groups surrounding the core atoms and two disulphide ligands between each two metal centres, bearing one aromatic ring each one. IR spectrum showed a different and complex pattern in the carbonyl region as compared to the parent and to the related compound **1** and it might be as a consequence of the asymmetrical distribution of the nine carbonyl groups. ^{1}HNMR spectrum showed one singlet correspondent to the four protons of the two aromatic rings. Elemental analysis suggests the molecular formula $Fe_4(CO)_9(\mu\text{-}S_2C_6H_2Cl_2)_2$ and MS FAB$^+$ experiments indicate the sequential loss of carbonyl ligands in addition to concordant theoretical and experimental isotopic distributions [13]. Essays to obtain a crystal are in process to determinate its structure. This complex **3** which has a structure that resembles the compound **2**, but doubled itself, could be employed in pyrolysis experiments and it is possible that might lead to CNM structures mainly due to the higher metal nuclearity and molecular mass of the cluster which could avoid the previous volatilisation problems, and furthermore, it is also possible that due to the odd number of carbonyl ligands its reactivity towards the chlorination process will be favoured. Preparation of these experiments is currently in process.

4. Conclusion

Analogous organometallic complexes containing metal centres, disulphide and carbonyl ligands, and an aromatic ring, were employed in synthesis of CNMs using two different approaches. The first compound was employed in bulk in CVD process and gave spherical metal nanoparticles of 20-50 nm in size surrounded by carbon matrix, and amorphous carbonaceous material. Using organometallic precursor in bulk showed very different performance as compared when deposited in silica, which suggests a searching for a further alternative substrate or a simpler deposition method. Second complex also used in bulk and pyrolysed at two different temperatures led to two extreme results in the chlorination process: at the higher temperature the precursor underwent an entire volatilisation, and at the lower one there was no reaction. In spite of this, it is possible that running experiments at intermediate temperatures might lead to CNMs, which are dependent of many variables. Moreover, a new compound was obtained and characterised, and it seems to be suitable as CNMs precursor.

Acknowledgements
JIA is grateful to Fundación Carolina (Spain) for a MChem scholarship. Prof F Zamora (Universidad Autónoma de Madrid) and Prof C Otero (Universidad Complutense de Madrid) are thanked for equipment facilities and TEM experiments.

References

[1] Ozin G A and Arsenault A C 2005 *Nanochemistry: A Chemical Approach to Nanomaterials* (Cambridge, UK: RSC Publishing)
[2] Nyamori V O, Mhlanga S D and Coville N J 2008 *J. Organomet. Chem.* **693** 2205-22
[3] Govindaraj A and Rao C N R 2002 *Pure Appl. Chem.* **74** 1571-80
[4] Coville N J, Mhlanga S D, Nxumalo E N and Shaikjee A 2011 *S. Afr. J. Sci.* **107** 1-15
[5] Vargas-Garcia J R and Goto T 2011 *IOP Conf. Ser.: Mater. Sci. Eng.* **20** 012001
[6] Urones-Garrote E, Ávila-Brande D, Ayape-Katcho N, Gómez-Herrero A, Landa-Cánovas A R and Otero-Díaz L C 2005 *Carbon* **43** 978-85
[7] Chen J, Vannucci A K, Mebi C A, Okumura N, Borowski S C, Swenson M, Lockett L T, Evans D H, Glass R S and Lichtenberger D L 2010 *Organometallics* **29** 5330-40
[8] Schwartz L, Singh P S, Eriksson L, Lomoth R and Ott S 2008 *C. R. Chim.* **11** 875-89
[9] Gonzalez-Prieto R, Nievas A, Fernandez-Vindel M A, Castillo O, Hernandez E, Delgado E and Zamora F 2011 *Dalton Trans.* **40** 3109-11
[10] Nyamori V O, Nxumalo E N and Coville N J 2009 *J. Organomet. Chem.* **694** 2222-7
[11] Mohlala M S, Liu X-Y, Witcomb M J and Coville N J 2007 *Appl. Organomet. Chem.* **21** 275-80
[12] Brauer G 1965 *Handbook of preparative inorganic chemistry* (New York: Academic Press)
[13] Ayuso J I 2011 *MChem thesis* (Madrid: Universidad de Alcalá, Universidad Autónoma de Madrid)

IOP Conf. Series: Materials Science and Engineering **45** (2013) 012002 doi:10.1088/1757-899X/45/1/012002

Iron removal from a kaolinitic clay by leaching to obtain high whiteness index

* R A Hernández Hernández[1], F Legorreta García[1], L E Hernández Cruz[1], A Martínez Luévanos[2].
[1]Universidad Autónoma del Estado de Hidalgo, Área Académica de Ciencias de la Tierra y Materiales.
Carretera Pachuca – Tulancingo, Km 4.5 s/n, Mineral de la Reforma, Hgo., México., C.P. 42184.
[2]Universidad Autónoma de Coahuila
Blvd. V. Carranza y González Lobo s/n; Saltillo, Coah., México. C. P. 25280
*Email: angelitofox3@hotmail.com

Abstract

Beige kaolin clays were crushed and sieved. The mineral was characterized by X-ray diffraction and chemically analyzed by atomic absorption spectrophotometry. The material below 37 microns was leached with oxalic acid. A study of main variables of the leaching process in different ranges was realized: time (0 to 120 min), temperature (25 to 100 °C) and acid concentration (0.01 to 0.5 M). Crystallographic analysis of the initial sample indicates the presence of kaolinite, iron oxides in the form of magnetite and iron-titanium oxides. By this method it was possible to obtain iron extraction percentages over 80% and whiteness index above to 90%.
Keywords: clays, kaolin, leached, oxalic acid.

1. Introduction

At present, kaolin minerals is a commonly used in the ceramics industry as raw material in various applications because of its refractory cement, glass fiber, etc.. Also is widely used in other industries such as paint, paper, pesticides, pharmaceuticals and cosmetics (1).

Many parameters influence the industrial use of kaolin especially its impurities as these largely determine the quality of kaolinitic minerals, affecting negatively their thermal and optical properties (2, 3, 4). For example, amounts as low as 0.4% iron (ferric) may be sufficient to impart color to the clay deposits and reducing their refractoriness and whiteness. Some specific physical and chemical properties of kaolin depend on its geographical source and methods of their processing, (1, 6), within which is the leaching, which is based on the removal or reduction of iron oxides, resulting concentrates that can be competitive kaolinitic up with finer kaolin, of higher purity than the kaolins Chinese (1, 5, 6). The leaching of iron from the red clay of kaolin is important for the following reasons: (a) You can find application in the production of refractory materials, (b) it is possible to recover items such as nickel, cobalt and chromium, (c) can be exploited to produce alumina as the Bayer process, or to produce anhydrous aluminum chloride (7).

Some previous studies have reported the leaching of kaolin with various chemical reagents, such as oxalic acid and other organic acids (7, 8, 9), in the presence of a fermented medium, bioleaching (4, 5, 10), etc., And all them in order to reduce the iron content. Even in some it has been observed that has received considerable attention mechanism studies with synthetic minerals such as hematite, goethite and magnetite ignoring the complex interactions that can be set in the dissolution of iron from industrial minerals, as bauxite, kaolin, silica and feldspathic sands (8).

The Hidalgo state in the municipality owns the Agua Blanca Iturbide there is a large deposit that is not exploited or benefit properly, for these reasons it is necessary to provide research results based on an increase in the quality (purity) of mineral giving it an added value.

Therefore, it is considered that the data obtained from the leaching of iron from oxalic acid kaolinitic clays are important for the diversity of its potential industrial applications, which is why this paper presents the study of kaolin ore leaching with in order to obtain the experimental conditions such as temperature and concentration of oxalic acid, leading to a high mineral white kaolin with low iron content.

IOP Conf. Series: Materials Science and Engineering **45** (2013) 012002 doi:10.1088/1757-899X/45/1/012002

2.- Material and methods

2.1 Material

The kaolin simple was (samples of particle size averaging 35μm) collected in the municipality of Agua Blanca de Iturbide, Hidalgo (México), with iron oxide content of 0.70% and a whiteness index of 87.80%, which was previously reduced in size and classified by ASTM mesh. In order to know the mineralogical composition, chemical and whiteness index of the samples of kaolin were used techniques X-ray diffraction (INEL mod. Equinox 2000), atomic absorption spectrophotometry (Perkin Elmer Co., Mod 2380) and reflectance spectrophotometer (GretagMacbeth, eye color model XTS).

2.2 Experimental procedure

The leaching tests were carried out in a round bottomed flask (500 mL) which was kept stirred in a heating mantle, For each run, 400 mL of oxalic acid solution ($C_2H_2O_4$, reagent grade) at different concentrations (0.01, 0.10 and 0.5 M) were added to the flask and the temperature was set to the desired value. Then, 40.0 g of clay was added to the flask while under 600 rpm magnetic stirring. Periodically (5, 15, 30, 60, 90 and 120 min), a 10 mL sample was taken from the leach slurry and was centrifuged immediately at 300 rpm during 15 min. A clear 5 mL aliquot of the solution was collected for total iron determination. All leaching tests were made at atmospheric pressure. In all experiments used a solid-liquid ratio of 10 g of kaolin and 100 mL of leaching solution. Experiments were done in duplicate. The variables studied were the concentration of oxalic acid (0.01, 0.10 and 0.50 M) and temperature (25, 40, 60 and 100 ° C).

3.- Results and discussion

The results of the mineralogical and chemical analyzes are shown in Figure 1 and Table 1, respectively. Was identified that the kaolinite (JCPDF 089-6538) and silica in the form of quartz (065-0466) and tridymite (076-0894), are the main constituents. The iron is considered as an impurity and is present as magnetite (089-0951), hematite (089-2810), Di-iron titanium oxides (074-2084) and greigeite (089-1998).

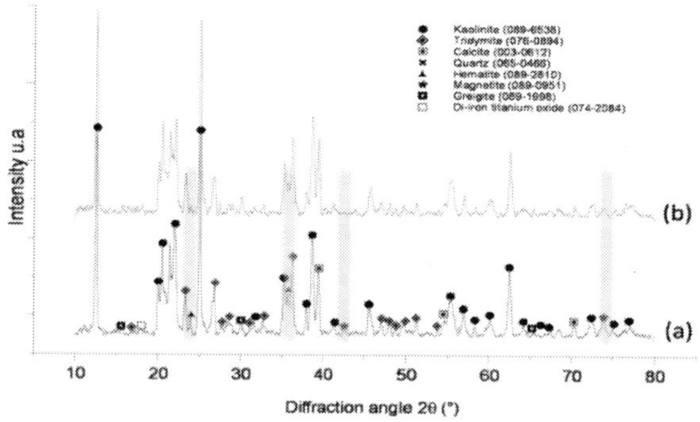

Figure 1. X-ray diffractogram of: a) kaolin ore with average particle size 35 μm. b) Leached kaolin.

Table 1. Chemical analysis of the sample of kaolin.

Components	Wt %
SiO2	65.00
Al2O3	27.00
Fe2O3	0.70
CaO	0.04
MgO	0.57
Na2O	2.65
K2O	2.42
TiO2	0.67
LIO*	0.95

LIO* = Loss on ingnition

It is important to make this identification because in several studies (2, 7, 8, 9, 11) has reported the dissolution of magnetite and greigite using oxalic acid. In this regard, Mandal, S. K. and Banerjee (13), reported that the dissolution of the hematite is much slower than for the other magnetite and hydrated iron oxides such as goethite (α-FeOOH) and lepidocrocite (γ-FeOOH); and suggest the use of different chemicals and experimental conditions to explain the mechanism of dissolution of hematite and magnetite. Many researchers (2, 7, 8, 10) have shown that this acid is most efficient for the dissolution of iron in clay minerals. Moreover, being an organic compound is not aggressive to the environment. Fig. 1 shows the decrease of the peaks corresponding to the reflections of magnetite and hematite for (a) ore kaolin and (b) leached kaolin.

IOP Conf. Series: Materials Science and Engineering **45** (2013) 012002 doi:10.1088/1757-899X/45/1/012002

In solution, oxalic acid ($H_2C_2O_4$) dissociates releasing the bi-oxalate ion ($HC_2O_4^-$):

$H_2C_2O_4 \rightarrow H^+ + HC_2O_4^-$ [1]

For its part, the bi-oxalate ion formed dissociates releasing the oxalate ion ($C_2O_4^{2-}$):

$HC_2O_4^- \rightarrow H^+ + C_2O_4^{2-}$ [2]

Of these species, it is considered that the bi-oxalate is responsible for the dissolution of the iron:

$Fe_2O_3 + H^+ + 5HC_2O_4^- \rightarrow 2Fe(C_2O_4)_2^{2-} + 3H_2O + 2CO_2$ [3]

Therefore, the conditions in the medium should be suitable leachant for favoring the prevalence $HC_2O_4^-$ ion. Reason it is important to evaluate the temperature and concentration of oxalic acid.

3.1 Effect of acid concentration

Figure 2 shows the dissolution of iron (%) versus time at 100 ° C, using various concentrations of oxalic acid (0.01, 0.10 and 0.50 M), is observed that the dissolution of iron is increased as the concentration of oxalic acid in the range 0 to 120 minutes. Obtaining a maximum rate of dissolution of 98 % at a time of 2 hours. This behavior is attributed to increasing the acid concentration increases the concentration of oxalate and hydrogen ions. That is, with the increase in the amount of oxalate ion favors the formation of bi-oxalate (see reaction 2), which is the species responsible for the extraction of iron and increased the hydrogen ions which has a direct influence on the pH. That coincides with that reported by Ambikadevi and Lalithambika in 2000 that found that oxalic acid was the best acid for the dissolution of iron (which unlike this work was in the form of goethite and hematite) of a kaolin mineral and to increase the acid concentration (0.05 to 0.15) increases the efficiency of iron dissolution (2).

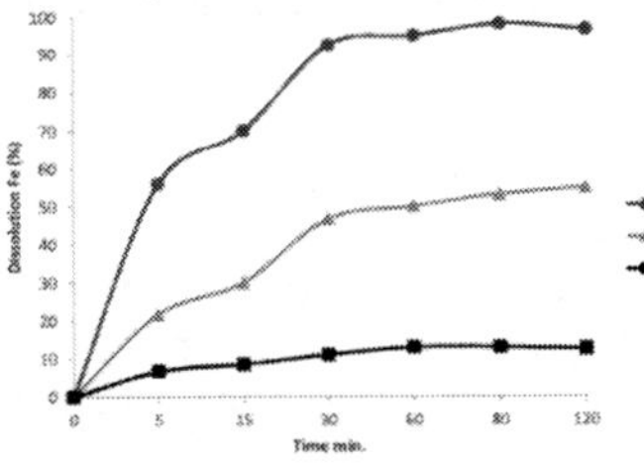

Figure 2. Iron extraction at 100 °C versus time at different concentrations of oxalic acid.

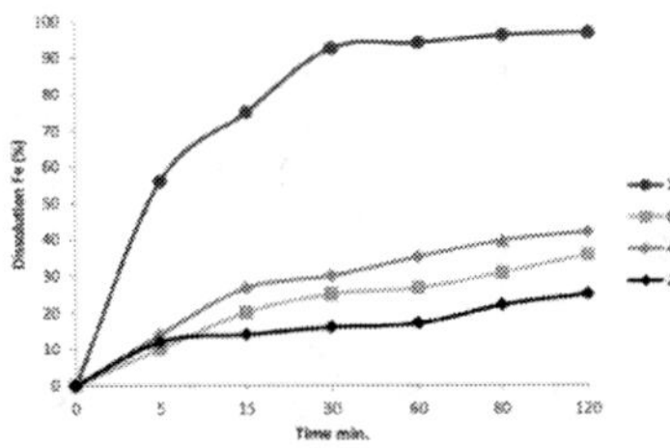

Figure 3. Iron extraction with oxalic acid [0.50 M] versus time at different temperatures.

3.2 Effect of reaction temperature

In Figure 3 presents the percentage of iron dissolution versus time, at a concentration of 0.5 M oxalic acid at various temperatures (25, 40, 60 and 100 ° C). It is observed that the extraction of iron increases with increasing temperature in the range of 0 to 120 minutes. That is, the dissolution of iron with oxalic acid has to be thermally activated to be efficient. This, on the other hand, is an indicator that the system exhibits a chemical control. Similar results for kaolinitic minerals have been reported by several authors as Mandal and Banerjee (13) and Lee et al, (11, 12). Figure 4 shows the results of the percentage of iron removed, and the ratio or percentage of whiteness obtained at a concentration of 0.50 M oxalic acid at 100 ° C. Can see that the whiteness index increases in the same way that increases the dissolution of iron ore, yielding a maximum whiteness index of 93.55% at a time of 2 hours. Confirming that a lower percentage of iron oxides, the whiteness index increases.

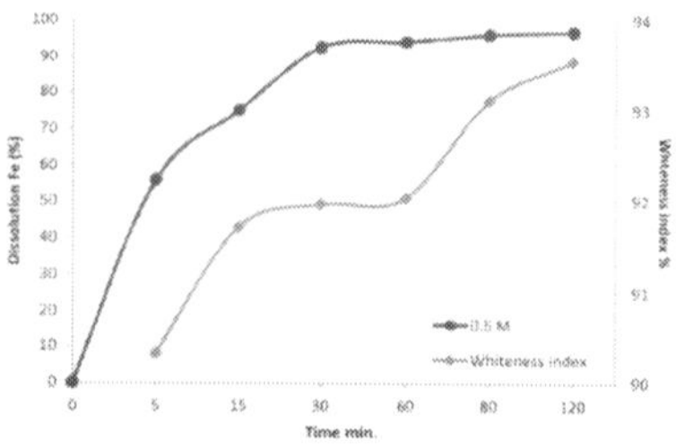

Figure 4. Iron extraction and whiteness index obtained at a concentration of 0.50 M oxalic acid at 100 ° C.

4.- Conclusions

This study found that using oxalic acid as leaching agent at a concentration 0.50 M and 100 °C in a time of 2 hours was possible bleaching kaolin clays from the municipality of Agua Blanca de Iturbide, Hidalgo (Mexico), obtaining iron extraction percentages of 98% and a whiteness index of 93.55%. Oxalic acid concentration and temperature markedly influenced the rates of dissolution and the whiteness index of the mineral studied.

References

[1]. Murray, H. H. 2000. "Traditional and new applications for kaolin, smectite, and palygorskite: a general overview", Applied Clay Science., vol. 17, pp. 207-221.

[2]. Ambikadevi., V.R., Lalithambika., M. 2000. "Effect of organic acids on ferric iron removal from iron-stained kaolinite", Applied Clay Science., vol.16, pp. 133-145.

[3]. Chandrasekhar, Sathy, Ramaswamy, S., 2002. "Influence of mineral impurities on the properties of kaolin and its thermally treated products", Applied Clay Science., Vol. 21, No.3-4, pp. 133.142.

[4]. Eun-You L., Kyung-Suk C., 1999. "Microbial removal of Fe (III) impurities from clay using dissimilatory iron reducers", Bioscience and bioengineering, Vol. 87, No.3, pp. 397-399.

[5]. Lee Eun-Young, Cho Kyung-Suk., Ryu Hee Wook. 2002. "Microbial Refinement of Kaolin by iron-reducing bacteria". Applied Clay Science, Vol. 22, pp. 47-53.

[6]. Saikia, N.J., Bharali, D.J., Sengupta, P., Bordoloi, D., Goswamee, R.L., Saikia, P.C., Borthakur, P.C., 2003. "Characterization, beneficiation and utilization of a kaolinite clay from Assam, India", Applied Clay Science, Vol. 24, pp. 93-103.

[7]. Veglio F., Toro L., 1994. "Process development of kaolin pressure bleaching using carbohydrates in acid media". Mineral processing, Vol. 41, pp. 239-255.

[8]. Veglio, F., Passariello, B., Toro, L., Marabini, A. M., 1996. "Development of a bleaching process for a kaolin of industrial interest by oxalic, ascorbic and sulphuric acids; preliminary study using statistical methods of experimental design". Ind. Eng. Chem. Res., Vol. 35, pp. 1680-1687.

[9]. Groudev, S. N., 1999. "Biobeneficiation of mineral raw materials". Miner Metall. Process., Vol. 16, no.4, pp 19-28.

[10]. De Mesquita, L.M.S., Rodríguez, T., Gomes, S.S, 1999. "Bleaching of brazilian kaolins using organic acids and fermented médium". Miner. Eng., Vol. 9, No.9, pp. 965-971.

[11]. Sung Oh Lee, Tam Tran, Byoung Hi Jung, Seong Jun Kim, Myong Jun Kim 2007. "Dissolution of iron oxide using oxalic acid". Science Direct Hydrometallurgy, No. 87, pp 91-99.

[12]. Ayala Perez G., Z., Vargas Rodríguez Y., C., Córdoba Tuta E., M., 2010. "Beneficio de una arcilla caolinítica de la región de Barichara (Santander) para la fabricación de refractarios". Dyna No. 164, pp. 29-38.

[13]. Mandal, S. K. and Banerjee, P. C. 2004. "Iron leaching from China clay with oxalic acid: effect of different physicochemical parameters", Int. J. Miner. Process, Vol. 74, pp, 263-270.

IOP Conf. Series: Materials Science and Engineering **45** (2013) 012003 doi:10.1088/1757-899X/45/1/012003

Development of process technologies for improvement of electroless nickel coatings properties

A Barba- Pingarrón[1,5], A Bolarín-Miró[2], F Sánchez - de Jesús[2], L Vargas-Mendoza[3], M Trujillo-Barragán1, P Molera-Sola[4], M A Hernandez-Gallegos[1] and R Valdez-Navarro[1].

[1] Centro de Ingeniería de Superficies y Acabados (CENISA). Departamento de Manufactura y Materiales. División de Ingeniería Mecánica e Industrial. Facultad de Ingeniería. Universidad Nacional Autónoma de México. 2° Nivel. Edificio "T" (Bernardo Quintana). Circuito Exterior. Ciudad Universitaria. Coyoacan, 04510, D.F. arbapin5@gmail.com.

[2] Cuerpo Académico de Sólidos Particulados. Instituto de Ciencias Básicas e Ingeniería. Universidad Autónoma del Estado de Hidalgo. México

[3] Departamento de Ingeniería Mecátrónica. Instituto Tecnológico y de Estudios Superiores de Monterrey. Campus Estado de México.

[4] Departamento de Ciencia de Materiales y Metalurgia. Universidad de Barcelona, España.

Email: arbapin5@gmail.com, abolarin@uaeh.edu.mx, fsanchez@uaeh.edu.mx, lvargas@itesm.mx, trujbar@gmail.com, pmolera@ub.edu, mahg22@yahoo.com,

Abstract. This paper describes research and technology developments that enable to improve nickel electroless coating properties. This work deals with: (a) different methods in order to achieve Ni-P-Mo coatings. (b) Other development is related with coatings with addition of hard particles such as SiC, WC or Al_2O_3,(c) Electroless nickel deposits on PBT and austempered ductile iron (ADI). (d) In addition, nickel coatings were deposited on powder metallic pieces and finally, electroless nickel coatings, in conjunction with layers from thermal spray process were formed. Characterization of all coatings by means of optical microscopy, scanning electron microscopy, micro-hardness, wear and corrosion tests were carried out. Results indicate positive increment in both mechanical and electrochemical properties which enhance field applications in Mexican industry.

1. Introduction.

Electroless (chemical) nickel coatings have been developed in the last 50 years and since then different applications in industrial fields have been reported [1]. Electroless nickel coatings offer interesting features such as: high-versatility of deposition on distinct materials (metals, polymers and ceramics), homogeneous and reproducible thicknesses (no matter shape or geometry of samples),

[5] To whom any correspondence should be addressed

amorphous or crystalline structures deposited as a result of variations in chemical composition baths, availability of heat treatments of coatings in order to achieve enhancement in wear properties and better. Electroless nickel coatings are mainly constituted by combinations of Ni-P at different mixture ratios. These formulations can be obtained either in acidic conditions or in alkaline environment from room temperature up to 100°C [1]. The purpose of this work is to study different routes deposition processes in order to improve mechanical and electrochemical properties which lead further applications.

2. Surface technology treatments

2.1 Development of ternary electroless coatings Ni-P-Mo.

Incorporation of molybdenum to electroless Ni-P coating achieved better wear resistance properties. The Ni-P-Mo bath was alkaline (pH ~9.5), with different percentages of P (9-11%) and Mo (3-8%) using a formulation reported in [2]. Conditions of the deposition were 88°C and 5 hrs. Some coatings, on low carbon steel, were subjected to heat treatment at 300°C, 400°C and 500°C from 30 to 240 minutes. Results show that amorphous coatings change their structure to crystalline as a consequence of heat treatments. Different percentages of crystallinity were observed by Philips XL 20scanning electron microscope (SEM). SEM, in conjunction with EDAX, X-ray diffraction in Siemens 530 Diffractometer, electrochemical tests in HCl solutions , wear measurements (pin on disc) and microhardness evaluations in Shimadzu equipment (100g load), were carried out on different samples. Results indicate that both crystallization and precipitation hardening processes induced a hardness increment from 600 VHN to 1450 VHN. The formation of grains, grain boundaries and the presence of precipitates of compounds Ni-P which are obstacles to dislocation movement explain this hardness improvement. In addition, wear resistance was also modified, with better properties, by similar reasons. However, corrosion resistance did not improve after surface treatments. The presence of grain boundaries, and compounds of Ni, P y Mo (Ni_3P, $MoNi_4$, for example), originates galvanic sites and reduces corrosion resistance. It has to point out that molybdenum incorporation induces instability in the baths provoking lower deposition rates; therefore, alternatives to solve this problem should be addressed. Figure 1shows typical morphology of these coatings.

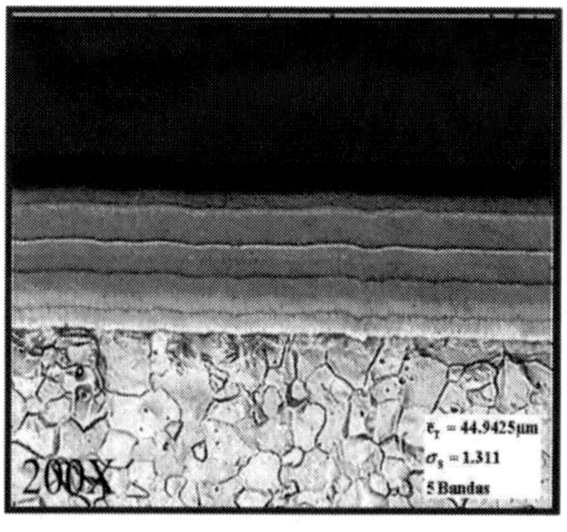

Figure 1. Image of Ni-P-Mo coating on AISI 1018 steel. OM.

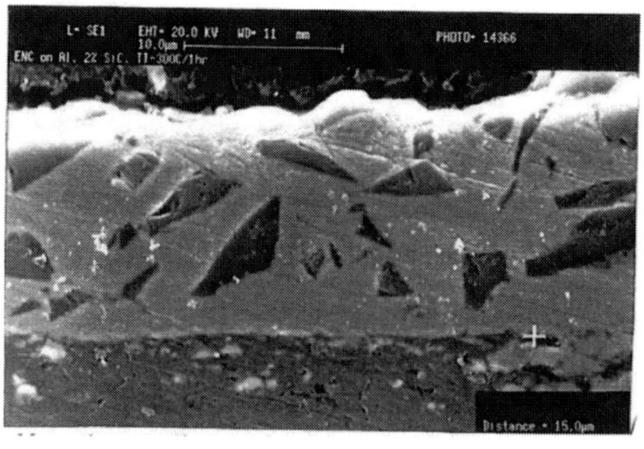

Figure 2. SEM micrograph of electroless Ni-P coating with SiC particles.

2.2 Electroless nickel plating with hard particles incorporated.

Incorporation of hard particles such as SiC, Al_2O_3 or WC into electroless nickel deposits on low carbon steel has been typical procedure studied for several years in CENISA. Proportion of particles used was 0.1 a 6 g/L, 85°C was the temperature and 4 hours the time, using mechanical or air agitation. The bath formulation employed has been reported in [3]. The main focus of incorporate hard particles into coatings relies on the fact that these aggregates induce better wear resistance, as indicated by Taber test results. That behavior is associated to the improvement of microhardness and rugosity generated by the presence of hard particles. In addition, it is also important to develop alternatives of stirring mechanisms of bath that enables homogenous dispersion into the metallic coating.

Concerning results of the characterization techniques described above, it is important to indicate the wear resistance increment. Figure 2 shows clear evidence of the obtained morphology aggregates of a electroless Ni-P coating.

2.3 Electroless Ni-P coatings on PBT

Polymer nickel-plated process represents a feasible chemical route to cover polymeric materials in a satisfactory way. This process involves long-term additional steps compared to traditional metal nickel-plating. One of the major challenges of cover polymers is the surface activation mechanism which directly affects the adherence properties. Results of the following work open a new and innovative trend of PBT metalizing since no references in literature can be found. The PBT nickel-plating procedure consisted of subsequent cleaning steps in conjunction with chemical attack using a solution containing chromic and sulfuric acid, surface sensitization with a $SnCl_2$ solution, activation mechanism using palladium chloride and finally electroless nickel-plating deposition employing a bath reported in [4] which it is usually carried before of electrolytic nickel plating. Figure 3 obtained in Philips XL20 SEM, shows a representative sample of Ni-P coating on PBT surface. The polymer surface that has been coated shows a uniform thickness and brilliant aspect.

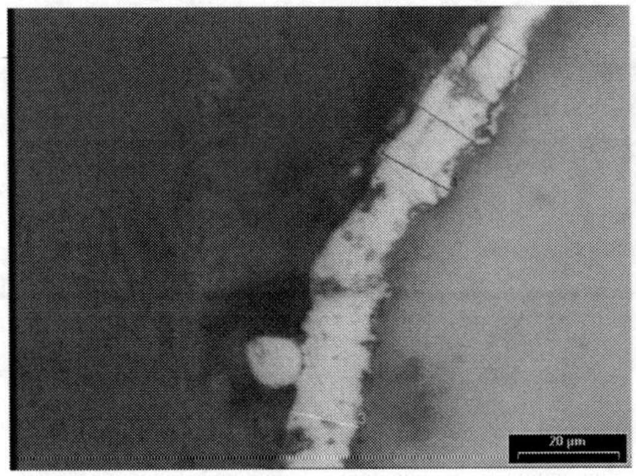

Figure 3. Scanning electron micrograph of a chemical nickel layer on PBT. (1000 X)

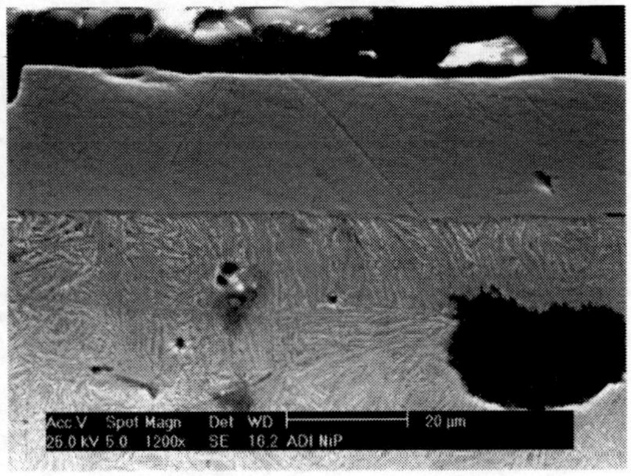

Figure 4. SEM micrograph of a Ni-P coating on austempered ductile iron (ADI). 1200 X.

2.4 Electroless nickel-plating on Austempered Ductile Iron (ADI)

ADI represents a recent engineering material with many industrial applications due to its attractive and adequate mechanical properties similar or even better to those of some alloy steels. Some applications require relatively low-cost material with good wear resistance for machinery devices as shafts, gears, and bearings and ADI is an attractive option [5]. ADI properties results from heat treatment cycles with final austempering process between 250 to 400°C. Metalizing ADI alloys represents a major challenge since no surface modifications are required after austempering process. Therefore, the present work developed a non-perturbation procedure for cover ADI with Ni-P alloys, using an acid bath formulation (pH 4.3 – 4.6) reported in [6] with a temperature of 85°C and 4 hours. Results show better corrosion properties after electrochemical tests in NaCl solution, without changes in the ADI microstructure. Figure 4, obtained in Philips XL20 SEM, depicts typical morphology of Ni-P coating deposited on ADI.

2.5 Duplex coating systems: Thermal Spray + Electroless nickel-plating on low carbon steel
Flame thermal spray process generates limited industrial coatings applications since morphology and defects of deposited material considerably affect adherence with irregular thicknesses along the sprayed surface. Therefore, the present work shows significant improvement in the corrosion properties of duplex coatings of thermal spray Ni-Cr-Fe-B-Si plus electroless Ni-P coatings. Prior to coating deposition chemical cleaning procedure was carried out in samples. Afterwards, thermal spray process was accomplished by standards procedures as reported elsewhere followed by an electroless nickel-plating. It was used an acid bath formulation (pH 4.3 a 4.6) [7], 85°C and 4 hours, after flame thermal spray procedure, employed operation parameters. This Ni-P layer acts as sealing barrier of defects of the previous coating, as can seen in electron micrograph obtained in Philips XL20 SEM (Figure 5). Therefore, improved adherence of the flame thermal sprayed layer in conjunction with final topcoat barrier provides an enhancement of corrosion properties, as indicated results of electrochemical tests in 3.5% sodium chloride solution.

Figure 5. Scanning electron micrograph of duplex flame thermal spray coating + electroless nickel coating on low carbon steel.

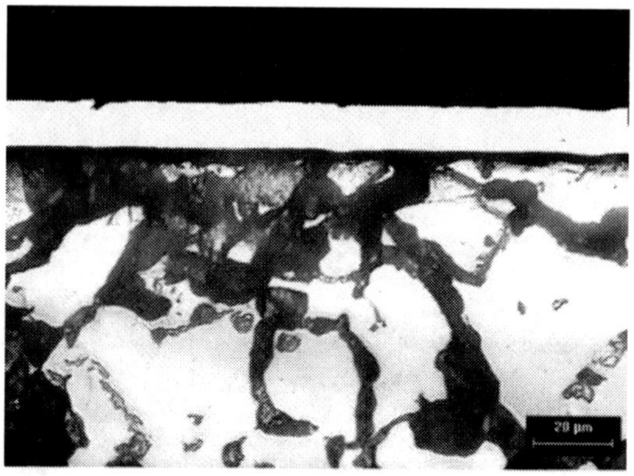

Figure 6. SEM micrograph of electroless Ni-P coating on iron substrate obtained by powder metallurgy.

2.6 Electroless nickel coatings on powder metallurgy materials.
Figure 6 shows a micrograph obtained in Philips XL20 SEM, of a electroless nickel coating deposited on an iron powder metallurgy sample using an acid bath [8]. Sponge and atomised iron compacts with different porosities were obtained by pressing at different densities (5.5 - 7.0 g cm^{-3}) and sintering at 1120 ° C in an N_2 - H_2 atmosphere. The ferroxyl test was used to evaluate the grade of superficial porosity, and the salt spray test helped to obtain the corrosion resistance of the Ni - P deposits on the PM compacts. Porosity of powder metallurgy samples represents disadvantages when a top coat layer is deposited. The present work determined that the amount of maximum porosity of iron powder

should be < 12 % in order to achieve good adherence between the samples and the electroless nickel coating. If samples generate elevated porosity poor coating deposition accompanied with low barrier and mechanical properties are achieved.

3. Conclusion

This work has presented different surface technologies alternatives developed in order to enhance the behavior of electroless nickel coatings. Deposition on PBT, ADI and powder metallurgy pieces show significance improvement in corrosion properties and wear resistance when topcoat layers of different Ni-P alloys are deposited. Hard particulates, ternary coatings alloys, heat treatments in conjunction with other surface technologies (thermal-spray), provide surface engineering systems that open new industrial applications.

4. References

[1] Barba A 2001 *Ciencia e Ingeniería de la Superficie de los Materiales Metálicos* ed. A J Vázquez and J J de Damborenea (Madrid: Textos Universitarios No. 31/CSIC) pp 261–279

[2] Vargas L, Barba A, Sánchez F and Bolarín A Age Hardening of Ni-P-Mo electroless deposit 2006 *Surf. Eng.* **22** 58

[3] Sánchez F, Bolarín A, Barba A, Torres and Hernández L Obtención y Caracterización de Recubrimientos Compuestos Ni-P-X (X = SiC y WC) vía química 2012 *Superficies y Vacío* **25** 128

[4] Trujillo M 2011 *Sobre el Niquelado Químico de Polímeros* (México: Ph D Thesis/UNAM)

[5] Nofal A and Jekova L Novel processing techniques and applications of austempered ductile iron (Review) 2009 *J. Univ. Chem. Tech. and Metallurgy* **44** 213

[6] Torres R 2007 *Obtención y Caracterización de fundiciones nodulares austemperizadas recubiertas por niquelado químico* (México: Sc. Master. Thesis/UNAM)

[7] Barba A, Hernández M, Sánchez F, Bolarín A and Valdez R 2011 *Obtención de Recubrimientos Duplex Níquel Químico – Proyección Térmica sobre Acero de Bajo Carbono. Efecto en su Comportamiento a la Corrosión en Medio Salino: Memorias del XVII Congreso de la SOMIM (San Luis Potosi, México, 21-23 September 2011)* p 719

[8] Bolarin A, Sanchez F, Barba A, Coreño O and Coreño J Electroless Nickel Plating of Atomised and Sponge Iron Compacts 2003 *Surf. Eng.* **19** 364

[9] Ploof L Electroless Nickel. Composite Coatings 2008 *Adv. Mat. Proc.* 36

[10] Nováka M, Vojtěcha D and Zelinkováa M Heat Treatment of Ni-P-Al_2O_3 Electroless Coatings 2009 *Metal* **19** 19

Acknowledgments

Authors (ABP, MAHG, MTB and RVN) acknowledge the support by DGAPA UNAM PAPIIT Research Project IT102612 "Desarrollo de Tecnologías de Modificación Superficial para la Optimización del Rendimiento de Materiales".

Sandblasting as a surface modification technique on titanium alloys for biomedical applications: abrasive particle behavior

J C Balza[1], D Zujur[1], L Gil[3], R Subero[3], E Dominguez[4], P Delvasto[1] and J Alvarez[2]

[1]Departamento de Ciencias de los Materiales. Universidad Simón Bolívar. Valle de Sartenejas, Caracas Venezuela.
[2] Departamento de Biomateriales. Sección de Salud. Fundación Instituto de Estudios Avanzados (IDEA). Valle de Sartenejas, Caracas. Venezuela.
[3]Universidad Nacional Experimental Politécnica "Antonio José de Sucre". Puerto Ordaz, Estado Bolívar, Venezuela.
[4]Departamento de Traumatologia del Hospital Uyapar. Puerto Ordaz. Venezuela.

E-mail: jbalzahernandez@gmail.com

Abstract. The present work shows the analysis of a sandblasting process using alumina abrasive particles on Ti-6Al-4V surfaces. The metallic samples were first characterized by optical microscopy (OM), revealing an α+β microstructure with a Widmanstätten morphology. Topography of the samples was assessed by scanning electron microscopy (SEM), before and after sandblasting. The Al_2O_3 particles used had a granulometric distribution between 420 and 850 µm, with a median particle size (d50) of 670 µm, which decreased to 420 µm after sandblasting for 10 seconds. This change in the size of the particles generated a loss on particle kinetic energy by a factor of 3.5. Such variation on processing conditions induced a progressive increase on average roughness (Ra) of the Ti-6Al-4V surfaces, until the first 7 seconds were reached. From that point on, a reverse process was observed, exerting a polishing effect on the surface of the Ti-6Al-4V alloy.

1. Introduction

Ti-6Al-4V is one of the most common alloys, which belongs to the (α+β) class of titanium alloys, due the presence of aluminum and vanadium. These alloying elements are used for phases' stabilization and for improving the mechanical properties and resistance to corrosion of the titanium alloy. It is well known that, for biomedical applications, Ti-6Al-4V is one of the most widely used alloys for orthopedic implants. However, it is important to create surface roughness to facilitate the process of osteointegration. There is not an optimum surface roughness, but some publications report an approximate value of 4 µm [1, 2, 3, 4, 5].

A very common technique to facilitate osteointegration through surface roughness is sandblasting with alumina particles (Al_2O_3). The size, shape, and kinetic energy of the particles, are important variables which influence the value of the surface roughness. In the sandblasting process, the particles are shooting as a consequence of the impulse provided by air pressure, allowing a gain of kinetic energy, which is directly proportional to the density, volume, and the square of the shooting velocity. The gain of kinetic energy is represented in the following equation

$$Ec = \rho.\left(\frac{2}{3}\right).\pi.r^3.V^2 \quad \text{Equation 1.1}$$

Where V, ρ, and r are the velocity, density, and particle radius, respectively.

Al_2O_3 has been considered one of the most commonly used abrasives for sandblasting due its hardness and acicular shape, as well as its low production cost with respect to other abrasive materials and its potential inert behavior in the osteoeointegration process. Nonetheless, fractures could occur during sandblasting process, as a consequence of the fragile nature of the alumina and the elevated kinetic energy of the particle stream at the time of impact.

The aim of the present work was to investigate the effect of the sandblasting treatment and the behavior of the Al_2O_3 particles in the surface modification of Ti-6Al-4V samples. This study allows the understanding of the surface properties of the Ti-6Al-4V alloy, which depend of the physical and mechanical characteristics of the employed abrasive, and also exert a great influence in the biological response of the human body.

2. Experimental Procedure

2.1. Alloy and metallographic analysis

Discs with 8 mm in diameter and 1 mm in thickness, were cut from the stem of a Ti-6Al-4V hip prosthesis. For their metallographic study, the samples were polished and attacked for 3 seconds with Kroll's solution (85 mL of H_2O, 5 mL of HNO_3 and 10 mL of HF), and their microstructure was then observed with an optical microscope (OLYMPUS PMG 3).

2.2. Sandblasting and determination of average roughness (Ra)

The surface of the titanium alloy' samples were modified using a sandblasting device. The pressurized air in the device, projects the abrasive Al_2O_3 particles at a 90° angle. The distance between the nozzle and the sample surface was 0.1 m. The air pressure was 0.3 MPa. For particle sizes between 420 – 600 µm, sandblasting was carried out for 2, 3, 4, 6, 7, and 10 seconds. At each sandblasting time, Ra was measured, employing a rugosimeter (Mitutoyo SJ-301) in three different parallel zones of the modified surface. Topography of the modified samples was analyzed by scanning electron microscopy (SEM) in a microscope (JEOL JSM-6390) with an acceleration voltage of 25 kV.

2.3. Granulometric analysis of the abrasive

Before and after sandblasting, the abrasive particles were studied by granulometric distribution, applying a sieving analysis from a representative sample of 1 kg, using a sieving machine (Gilson Inc. SS-15) for 10 min.

3. Results and Discussions

3.1. Alloy and metallographic analysis

Figure 3.1 shows a micrograph of the study alloy. This picture corresponds to a biphasic microstructure, where the lighter zone is the alpha phase and the dark zone represents the beta phase. The morphology is of the Widmanstätten. type, with results from the nucleation and growth during the transformation of alpha and beta, with the alpha phase growing on the preferential crystallographic planes within the beta phase matrix. In other words, the alpha phase nucleates on the edges of the beta matrix granule and grows in the form of metastable plates. The revealed morphology indicates that the alloy possesses great resistance to tension and fluency, due to the fact that the Widmanstätten plates help in improving the mechanical behavior.

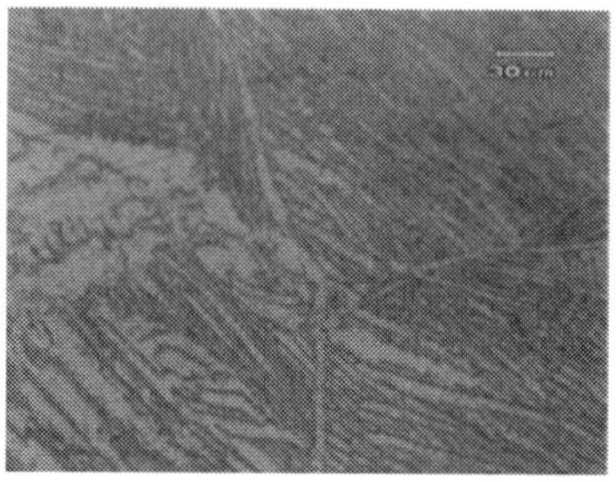

Figure 3.1. Micrograph of T-i6Al-4V alloy chemically attacked with Kroll's solution (200X)

3.2. Sandblasting process and determination of surface roughness (Ra)

Once the sandblasting process was carried out, the average Ra was measured on the metallic surfaces. The histogram presented in Figure 3.2 shows the values of Ra obtained as a function of sandblasting time, when the particle size was maintained in the range of 420-600 μm.

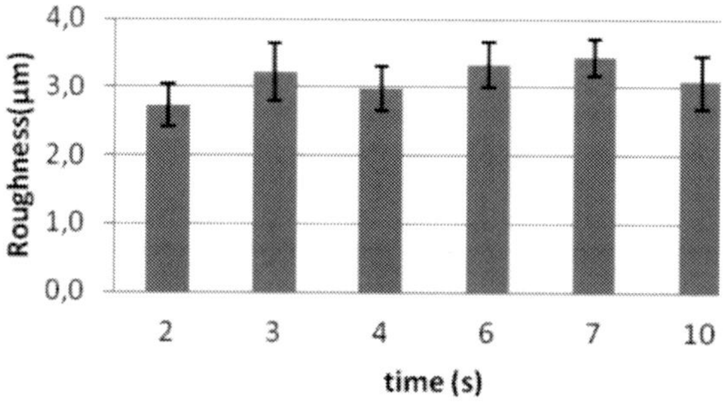

Figure 3.2. Histogram of average surface roughness of the Ti6Al4V samples (n = 33) as a function of sandblasting time, using a particle size in the range of 420- 600μm

In Figure 3.2, it can be observed that there is an optimum roughness value of 3.4 μm at 7 seconds of sandblasting ($p < 0.1$). However, this value decreases to 3.1 μm when the time is increased to 10 seconds. This phenomenon is possibly a consequence of the rupture of the alumina particles when impacting the surface.

The micrographs shown in Figure 3.3 correspond to Ti-6Al-4V surfaces before and after the sandblasting treatment. Figure 3.3a shows the unmodified surface, where only grinding marks caused during sample preparation can be appreciated. Figure 3.3b shows a substantial change in surface morphology, product of the plastic deformation associated to the impact of alumina particles.

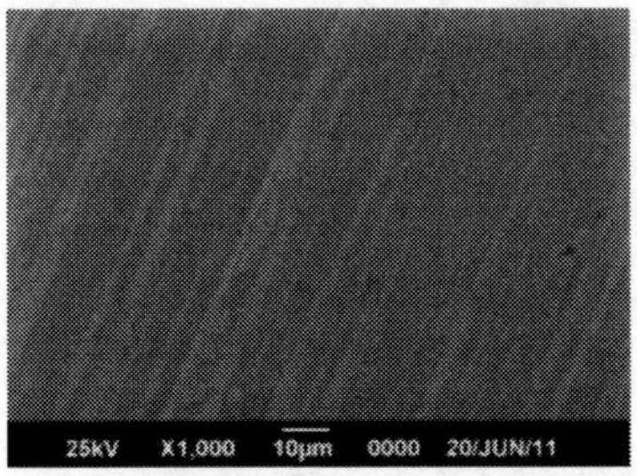

Figure 3.3a. Scanning electron micrograph of unmodified Ti6Al4V at 1000X.

Figure 3.3b. Scanning electron micrograph of sandblasted Ti6Al4V at 1000X.

3.3. Characterization of the abrasive particles

Figure 3.4 shows the results obtained from granulometric distributions of the abrasive particles, before and after the sandblasting process.

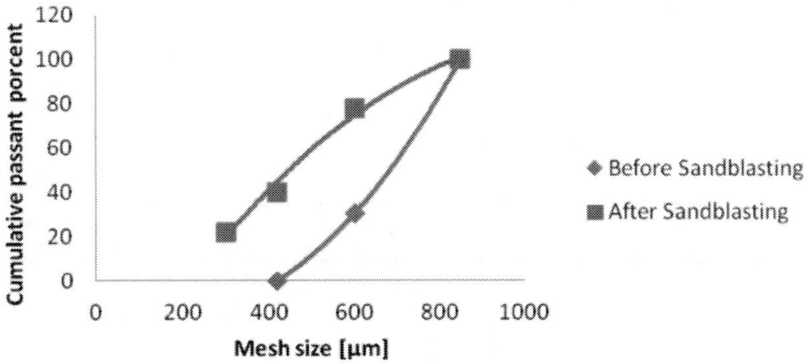

Figure 3.4. Granulometric cumulative curves showing d50 of the abrasive granulometric distributions before and after sandblasting.

Emphasizing on the granulometric curves in Figure 3.4, it is established that the medium particle size, d50, is 670 µm for a granulometric distribution before sandblasting process, and 450 µm after this process. This result indicates a clear decrease in the abrasive particle size. This fact is an empirical evidence of that the particle fracture is associated with the tendency to decrease roughness, with prolonged sandblasting times.

This phenomenon can be explained by analyzing the variation in kinetic energy of the particle stream. According to equation 1.1, the kinetic energy of the particles was reduced 3.5 times when the diameter changed from 670 μm to 420 μm. Therefore, the impact received by the Ti-6Al-4V surface was drastically reduced in time, affecting the plastic deformation of the surface. On the other hand, when the size decreases, it is also true that the number of particles projected increase in an inverse proportion to the decrease in kinetic energy. This can induce an inverse effect in surface treatment, resulting in a polishing finish. Thus, in the case of this study, the average surface roughness could decrease after 10 seconds of sandblasting, when compared to shorter processing times.

4. Conclusions

Alumina particles present ruptures during the sandblasting process, causing a significant drop of their kinetic energy. Consequently, smaller particle sizes during sandblasting generate a decrease on surface roughness of the Ti-6Al-4V samples.

References

[1] Le Gu´ehennec L., Soueidan A., Layrolle P. and Amouriq Y 2007 *J. Dent. Mater* **23** 844.
[2] Ranolda J., Lyngstadaasb P. and Ellingsena E 2003 *Biomaterials* **24** 4559.
[3] Isabel D, Risa I., Hiroqui K, Ken-Ichiro T., Naoko Y., Toshi-Ichiro T., Kuntteru N. and Masahiko M. 2009 *J. Gifu. Dent. Soc.* **36** 89.
[4] Deligianni D., Katsala N., Ladas S., Sotiropoulou D., Amedee J. and Missirlis Y.2001 *Biomaterials* **22** 241.
[5] Wennerberg A.and Albrektsson T. 2009 *Clin. Oral Impl. Res.* **20** (Suppl. 4) 172.
[6] Vander G. 1984 *Materials Science and Engineering Series* ed. McGraw-Hill pp 140-141, 250-253.
[7] ASM: ASM Metals Handbook 2004 *Metallography and Microstructures*. Vol 9.
[8] Benedetti M. and Fontanari V. 2004 *Fatigue Fract Engng Mater Struct.* **27** 1073.
[9] Leinenbach C. and Eifler D. 2006 *Biomaterials* **27** 1200.

IOP Conf. Series: Materials Science and Engineering **45** (2013) 012005 doi:10.1088/1757-899X/45/1/012005

Thermal behavior of poly(2-hydroxyethyl methacrylate-bis-[trimethoxysilylpropyl]amine) networks

L A Bustos Figueroa[1*], R Salgado Delgado[1*], E García Hernandez[1], Z Vargas Galarza[1], E Rubio Rosas[2] y R Salgado Rodriguez[3]

[1] Instituto Tecnológico de Zacatepec, División de Estudios de Posgrado e Investigación y el Departamento de Ingeniería Química y Bioquímica del Instituto Tecnológico de Zacatepec
[2] Benemérita Universidad Autónoma de Puebla, Centro Universitario de Vinculación de la Benemérita Universidad Autónoma de Puebla
[3] Instituto Tecnológico de Tijuana, Centro de Graduados e Investigación del Instituto Tecnológico de Tijuana

E-mail: renesalgado@hotmail.com

Abstract. Poly(HEMA-BisSi) networks were prepared by using acid-catalyzed sol-gel of bis-[trimethoxysilylpropyl]amine (BisSi) and free radical polymerization of 2-hydroxyethyl methacrylate (HEMA). The thermal properties of the poly(HEMA-BisSi) networks were investigated with differential scanning calorimetry (DSC) and thermogravimetric analysis (TGA). The thermal behavior of these networks was also compared with homopolymers (The networks formed in both PHEMA and PBisSi gels were identified). The glass transition temperature (T_g) of PHEMA homopolymer was found as 103.74 °C. The thermal degradation of the poly(HEMA-BisSi) networks with different silica contents (e.g. 10, 15 and 25 wt%) were evaluated with use of DTG. It was observed that the thermal degradation temperature of poly(HEMA-BisSi) networks changed much with the BisSi content.

1 Introduction

Poly(2-hidroxyethyl methacrylate)(PHEMA) have been widely used in the pharmaceutical industry because of biocompatibility and friendly characteristics[1-7]. It is interesting to study how the thermal stability of PHEMA of is affected by incorporating BisSi (bis-[trimethoxysilylpropyl] amine). It is interesting choice because its hydroxyl groups, in addition to forming hydrogen bonds will eventually form Si-O-C by condensation with silanol groups[8,9,15]. Consequently, design of efficient response active materials that are also thermally stable has been a long sought after goal [10].

The use of sol-gel method [11-13] form synthesis of new materials is particularly appealing because the low temperature solution-based synthesis route is amenable to molecular chemistry approaches such that one can modify the structure and function of the parent silica glass by structure modification of the precursor. One strategies involves the simultaneous growth of the inorganic phase via acid hydrolysis and condensation of alkoxysilane precursor and the polymer phase via free radical polymerization of vinyl monomer in a common solvent [8].

In this study, we report the thermal properties of poly(2-hydroxiethyl methacrylate-bis[trimethoxy-silylpropyl]amine) networks.

2 Experimental

Into in three-neck flask were placed BisSi, HEMA, THF and HCl, equipped with a condenser, a magnetic stirrer and a nitrogen gas inlet-outlet. Polymerizations were carried out in an inert gas atmosphere. In a typical procedure for the preparation of 15 wt% SiO_2 poly(HEMA-BisSi) hybrid material, 51.22 g (0.15 mol) of bis-[trimethoxysilyl- propyl]amine and 45.07g (0.346 mol) of HEMA were dissolved in 36.05g(0.50 mol) of THF at room temperature. Then, 1.250 g of 0.5 M HCl was added to solution. The solution was refluxed at 70 °C for 2.5 h. After the polymer solutions was cooled to room temperature, 0.0042 g (0.017 10^{-3} mol) of benzoyl peroxide (BPO) was added to the solution and dissolved. At room temperature, the solvent (i.e. THF) and by-products from sol-gel reactions (i.e. MeOH and water) were removed via vacuum distillation. The transparent solutions were finally dried for 12 h in an oven at 80°C to complete the polymerization. The resulting xerogels (powders) formed were comminuted in an agate mortar for subsequent analysis [8].

2.2 Materials

Bis-[trimethoxysilylpropyl]amine (BisSi, Aldrich) and 2-hydroxyethyl methacrylate (HEMA, Aldrich), tetrahydrofuran anhydrous (THF, Aldrich), benzoyl peroxide (BPO, Aldrich) and hydrochloric acid (HCl, J.T. Baker) were used without further purification.

2.3 Measurements

Thermalgravimetric analyses (TGA) were conducted (Netzsch, instrument STA449F3 Jupiter) by heating powdered samples to 800 °C at 10 °C with a flow of 20 mL N_2/min. The random errors of weight losses were determined for samples with selected compositions. Differential scanning calorimetry (DSC) traces were recorded (Netzsch, instrument STA449F3 Jupiter).

3 Results and discussion

The use of bis-[trimethoxysilylpropyl]amine (BisSi) as the sole component in a coating is limited, however, as gelation occurs soon after mixing with water [14]. The bis-alcoxysilan (BisSi) and HEMA (via free radical polymerization) after addition of HCl are likely to under hydrolysis and condensation reactions that occur either by elimination of methanol or water molecules. The hydrolysis of BisSi and HEMA (Scheme 1) may involve the following reaction steps:

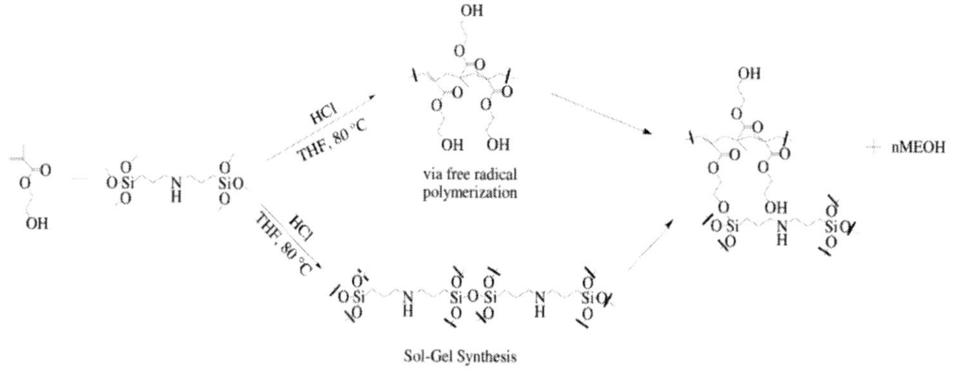

Scheme 1

3.1 Thermal analyses

3.1.1 DSC

DSC sensitively defects the small heat capacity change due to the glass transition temperature for PHEMA and the poly(HEMA-BisSi) networks (Figure 1). DSC measurements performed on the hybrid polymers revealed that did not have a significant glass transition temperature (T_g) and the T_g of the baseline polymer, PHEMA, at about 103.74°C. This lack of observable transition in the poly(HEMA-BisSi) samples is most likely attributable to the influence of the crosslinking that occurs via crosspolymerization of the growing PBisSi polymer chains with HEMA pendant groups on the silica structure formed by the condensation of PHEMA with the silica. Costa R. [15] and Li S. [8] reported effect over the behavior the hybrids materials in DSC similar to this results.

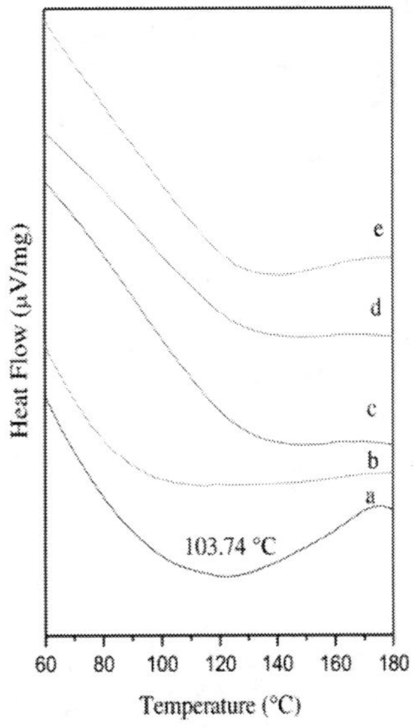

Figure 1 DSC thermograms of Poly(HEMA-BisSi) networks containing different content of BisSi (10, 15, 25 wt%) and homopolymers. (a)PHEMA, (b)PBisSi, (c) Poly(HEMA-BisSi)10%1.5, (d) Poly(HEMA-BisSi)15%1.5 and (e) Poly(HEMA-BisSi)25%1.5.

3.1.2 TGA

To examine the composition effect on thermal degradation poly(HEMA-BisSi) networks, we measured TG-curves for all the samples from 27 to 800 °C. The poly(HEMA-BisSi) hybrid polymer show improved thermal stability as observed by a reduction in mass loss rate from the TGA measurements in N_2 atmosphere (Figure 2). The results of TGA analyses show that the weight loss for all hybrids occurs in two steps. The first step is due to the evaporation of non-bonded water and other volatile compounds from room temperatures to ~200 °C [15]. The second step, is due polymer degradation and loss of formation a percolated poly(HEMA-BisSi) network [8]. From DTG curves, the maximum temperature of weight loss were also noted in Figure 3 the mass loss of PBisSi begins at 496.92 °C and at maximum temperature 517.23 °C, such as shown in the Figure 3. However, initial degradation temperatures of poly(HEMA-BisSi) networks were increased slightly in all samples compared with PHEMA and with increasing BisSi content. As can be seen Figure 2 poly(HEMA-BisSi) networks degrade in two steps. This is evidenced by appearance of distinct peaks in DTG thermograms. On the other hand, the mass loss of PHEMA reaches maximum at 428.71 °C. The curve of PHEMA homopolymer indicates one reaction stage (Figure 3) which is reflected as single peak in the DTG curve (Figure 3).

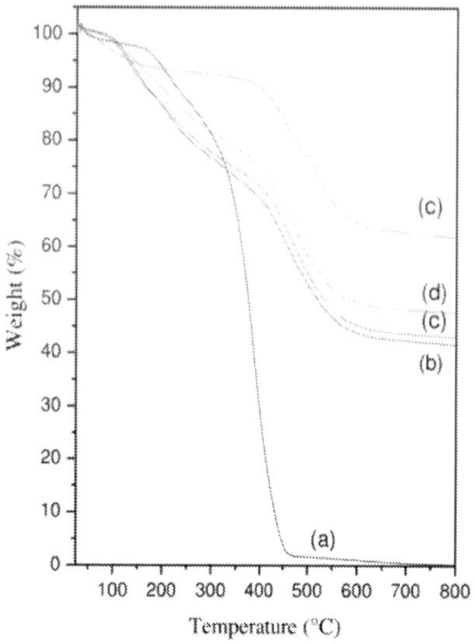

Figure 2 TGA curves of the hybrids samples and control samples recorded at a heating rate of 10 °C/min in N_2: (a) PHEMA; (b) PBisSi; (c) Poly(HEMA-BisSi)10%1.5; (d) Poly(HEMA-BisSi)15%1.5 and (e) Poly(HEMA-BisSi)25%1.5.

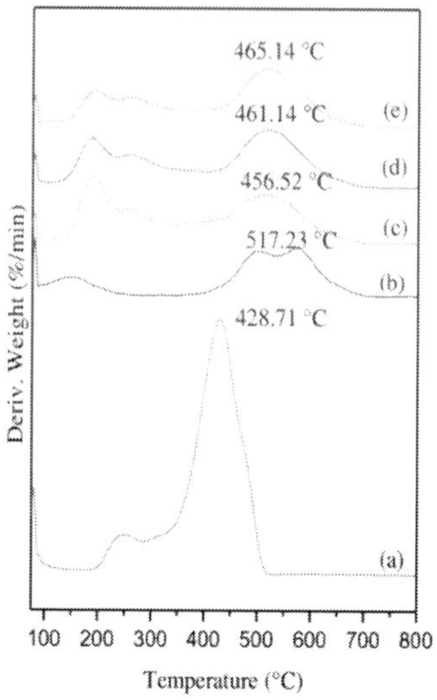

Figure 3 DTG thermograms of poly(HEMA-BisSi) networks containing different content of BisSi (10, 15, 25 wt%) and homopolymers. (a) PHEMA; (b) PBisSi; (c) Poly(HEMA-BisSi)10%1.5, (d) Poly(HEMA-BisSi)15%1.5 and (e) Poly(HEMA-BisSi)25%1.5.

4 Conclusions

The results clearly demonstrate that this approach would allow for the synthesis of organic-inorganic materials by conventional casting. Based on the TGA results and DTG observation of de stability of poly(HEMA-BisSi) networks, a stabilization model is proposed. The weakly acidic silanol group (-OH) on the silanol products and groups (-OH) on the PHEMA, forms hydrogen bonds with the strongly secondary amine group (-NH) on the PBisSi. On the other hand, T_g of the poly(HEMA-BisSi) networks cannot be determined with accuracy.

Acknowledgements

To BUAP-CUV and DGEST-4416.11-P for the financial support.

References

[1] Reining B, 2002 *Polymer* **43** 3139.
[2] Ma X and Wang H 2012 *International Journal Nanomedicine* **7** 1313-28.
[3] Juhasz J A 2010 *Sci. Thecnol. Adv. Matter.* **11** 1-7
[4] Bavarresco V P and Zavaglia C A C 2002 *Materials Research*, **5** 4481-84.
[5] Cretu A and Gattin R 2004 *Polymer Degradation and Stability* **83** 399-404.
[6] Santi M 1996 Biomaterials **17** 1459-67.
[7] Abraham S and Brahim S 2005 *Biomaterials* **26** 4767-78
[8] Li S 2007 *Polymer* **48**, 3982-89.

[9] Zhu D 2004 *Progress in Organic Coating* **49** 42-53.
[10] Caykara T 2003 *Polymer Degradation and Stability* **80** 339-43.
[11] Azolin D R 2004 *Journal of No-Cristaline Solids* **337** 201-6.
[12] Yano S 1998 *Materials Science and Engineering C* **6** 75-90.
[13] Kim Goo-Dae 1999 *Applied Organometallic Chemistry* **13** 361-72.
[14] Metroke T 2009 *J.Sol-Gel Sci Technol* **51** 23-31.
[15] Costa R O R, Vasconcelos W L. 2002 *Journal of Non-Cristalline,* **304** 84-91.

Nanograin formation in milled MoO_3 powders

J Guerrero-Paz[1], H Dorantes-Rosales[2], J A Aguilar-Martínez[3] and V Garibay-Febles[4]

[1]Ceramics Lab, AACTYM-ICBI-UAEH, Pachuca, Hidalgo, CP 42184, Mexico
[2]Transmission Electron Microscopy Lab, ESIQIE-IPN, D.F., CP 07300 Mexico
[3]X-Ray Lab, CIMAV, Apodaca, Nuevo León, CP 66600 Mexico
[4]Ultra High Resolution Electron Microscopy Lab, IMP, D.F. CP 07730 Mexico

E-mail: guerrero@uaeh.edu.mx

Abstract. Powder of Molybdenum trioxide was milled for different times in horizontal ball mills. Such powder was characterized by TEM and XRD. Powder was rapidly de-agglomerated and fragmented up to attain nanoplates of two types, amorphous and crystalline. Finally, cold-welding of nanoplates occurred permitting some relaxation process to obtain a more stable energized structure consisting of equiaxial crystalline nanograins after 16 hours of milling.

1. Introduction

It is known that a nanostructure in powders is currently obtained by ball milling. Phenomena as deformation, fracture and cold-welding occur, predominating one of them if powders are ductile or fragile [1]. The microstructural evolution during milling of fragile powders indicates that fragmentation and aggregation of particles predominate until reach a limit in particle size, after this, conditions are favorable for the cold welding or other material transfer mechanisms.

During milling of ductile powders, defects are created and rearranged, permitting the formation of a stable "energized" structure consisting of nanograins [2]. Understanding of how nanograins are attained in fragile materials by milling is not clear. By other hand, orthorhombic MoO_3 powder is a material that consists of agglomerates of monocrystalline plates bonded by Van der Waals forces [3], which if they were milled, probably showed a clear transition step to obtain an equiaxial nanograin structure. Mestl et al [4] followed mainly by XRD the microstructural evolution of milled MoO_3 powders, where they found that the grinding process seems to proceed via two steps: disintegration of larger particles or agglomerates and finally a continuous fragmentation of primary cristallites. A TEM study is needed to get a better understanding how nanograin is formed during milling of MoO_3 powders

2. Experimental

MoO_3 powder obtained by roasting of molybdenum concentrates (Molymex S.A. de C.V.) with a nominal purity of 99.44% and a mean particle size of 100 μm was used. Millings of powder were done at 0.5 h, 4 h, 8 h and 16 h in stainless steel mills of diameter of 12.5 cm. 2.461 Kg of hardened steel balls with sizes of 1.27 cm were utilized. Millings were carried out at 110 rpm for different times under air atmosphere. A ball-to-powder volume ratio of 90 was chosen as optimum for this powder after some previous experiments. An ultra high resolution transmission electron microscope (TEM) Jeol, JM2100 with a LaB_6 filament was used at 200 KV. Sigma ScanPro V 5.0 software was used for image analysis. An X-Ray diffractometer Philips, X'Pert, with a secondary graphite monochromator was employed to obtain XRD patterns with a step of 0.03^0 and a time per step of 1 second. Monochromatic Cu Kα_1 radiation was used.

3. Results and discussion

3.1 X-ray diffraction of MoO_3 powders milled for different times

It is seen in figure 1 that intensity of peaks of MoO_3 milled for 16 hours is smaller than that of the un-milled ones and a broadening of its width is also visible, which means that a nanograin structure was attained at that milling time. Impurities of MoO_2 and SiO_2 phases are present in un-milled powder.

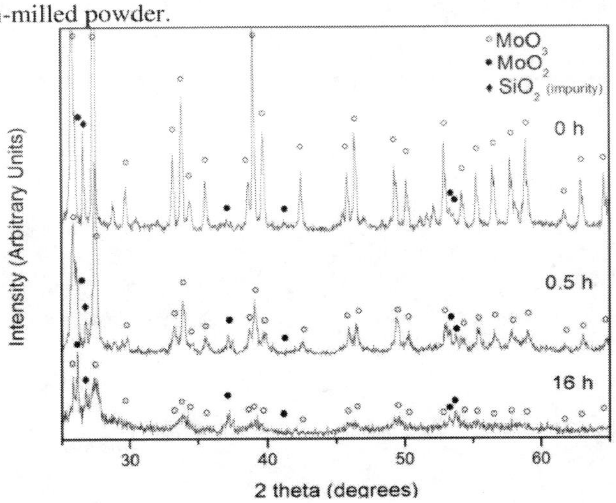

Figure 1. X-ray diffraction patterns of MoO_3 powder milled and un-milled (0 h). Impurities of MoO_2 and SiO_2 are detected. Powder Diffraction files from the ICDD used to identify MoO_3, MoO_2 and SiO_2 phases were 00-035-0609, 01-076-1807 and 00-083-0539 respectively. MoO_3 phase crystallizes in the orthorhombic lattice (space group *Pbnm (62)*)

3.1. De-agglomeration, fragmentation and cold-welding of monocrystalline plates

Mestl et al studied the milling of MoO_3 powder mainly by XRD [4]. They described the grinding process as a de-agglomeration and fragmentation of monocrystalline plates involving amorphization of their surface. TEM photomicrographs of figure 2 show these different steps in MoO_3 powders milled at 0.5 h, 4 h and 8 h in the present work.

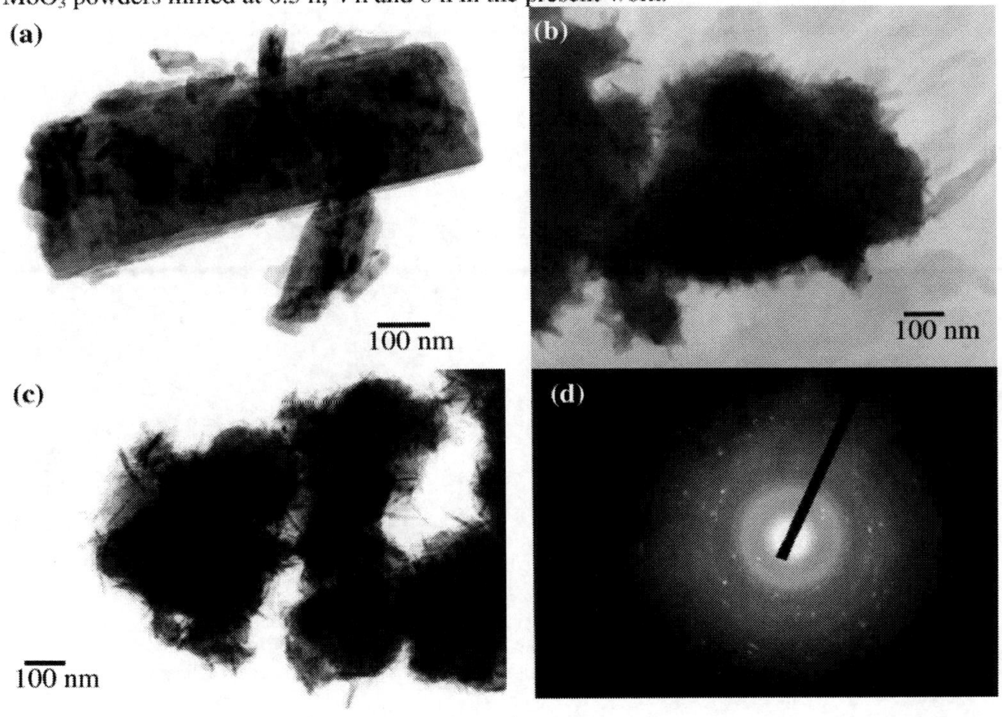

Figure 2. Bright field TEM images of MoO_3 powder milled: (a) 0.5 h, (b) 4 h and (c) 8 h. (d) SAD pattern of (b) showing a broader ring next to the central beam.

A continuous de-agglomeration and fragmentation of plates happen during the early milling times (see figure 2a) until reach some limit size at intermediate times (for example, 4 or 8 hours). After this, a possible exfoliation of nanoplates happens that liberates monolayers. Two types of monolayer nanoplates are observed, amorphous and crystalline ones. Also, at intermediate times of milling, cold-welding of nanoplates occurs, resulting in microparticles where individual nanoplates are still visible at its border (see figures 2b and 2c). The amorphous phase is detected by a broader ring in the selected area diffraction (SAD) pattern showed in figure 2d.

3.2. Limit size in Nanoplates and nanograins

High resolution TEM views show the amorphous or crystalline nature of the nanoplates. The amorphous ones are showed in figure 3a where some characteristic rounded zones at the interior of the nanoplates are clearly seen. A crystalline nanoplate is shown in figure 3b.

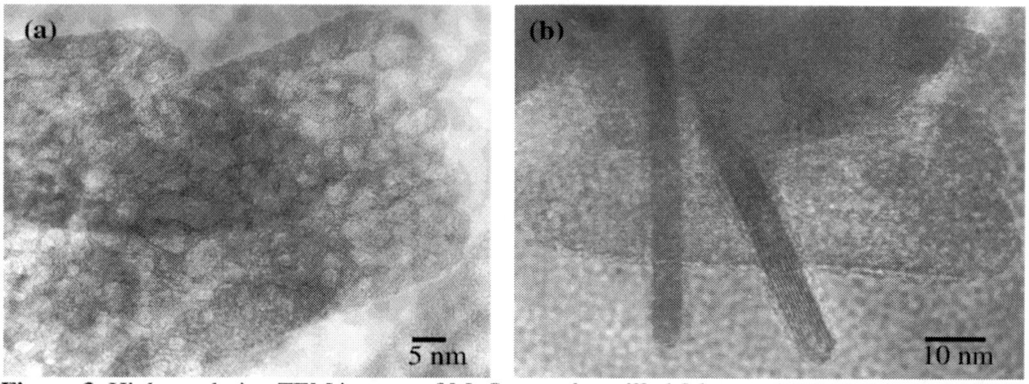

Figure 3. High resolution TEM images of MoO_3 powder milled 8 hours: (a) Amorphous and (b) crystalline nanoplates.

At long milling times (16 hours) microparticles are composed of only equiaxial crystalline nanograins as seen in figure 4. A complete disappearance of nanoplates suggests that some kind of recovery of crystalline nanoplates and recrystallization of the amorphous ones occur such that a lower-energy structure is formed as equiaxial crystalline grains. Internal energy that promotes such relaxation process must be stored at the surface of nanoplates but also in the disorder structure of the amorphous phase.

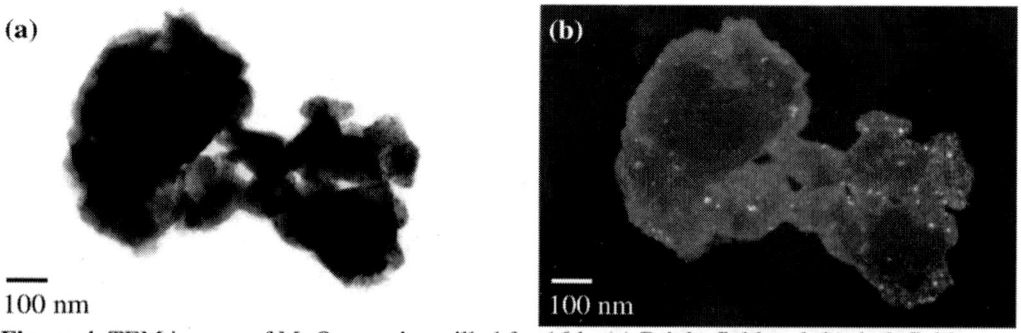

Figure 4. TEM images of MoO_3 powder milled for 16 h: (a) Bright field and (b) dark field images.

During the milling of MoO_3 powders a limit in size is attained for both, nanoplates and nanograins. Table 1 shows that amorphous nanoplates are wider than crystalline ones. The equiaxial grain is smaller than the width of crystalline nanoplates. Some dimensional factor, related to surface area could be influencing the transition of amorphous-to-crystalline phase in nanoplates to attain a minimum value of surface stress [5]. A study to explain the presence of amorphous and crystalline nanoplates is under current.

Table 1. Width of nanoplates and Feret Diameter of equiaxial nanograins.

	Nanoplates		
	Amorphous (nm)	Crystalline (nm)	Grain size (nm)
mean	16.13	6.09	3.81
Standard deviation	6.29	2.36	2.48
Sample size	23	68	351

4. Conclusions

A more complete description of the microstructural evolution of milled MoO_3 powders suggested by Mestl et al [4] has been done in the present work. Observations by TEM suggest that during milling of MoO_3 powders a rapid de-agglomeration and fragmentation of monocrystalline plates is happening at short milling times. A limit in size of nanoplates is attained mainly in its width during intermediate milling times. Some kind of exfoliation occurs of these nanoplates until monolayers are observed. Two types of monolayer nanoplates are seen, amorphous and crystalline ones. Also during intermediate milling times cold-welding of monolayer nanoplates predominates forming microparticles where individual monolayer amorphous and crystalline nanoplates are still observed at its border. At long milling times no more nanoplates are observed. Microparticles are composed of only equiaxial crystalline nanograins. Such fact suggests that a relaxation process occurred.

5. References

[1] Koch C 1989 *Annu. Rev. Mater. Sc.* **19** 121-43
[2] Koch C 1993 *Nanostructured Materials* **2** 109-29
[3] Anderson G and Magnéli A 1950 *Act. Chem. Scand.* **4** 793
[4] Mestl G Herzog B Schlögl and Knözinger H 1995 *Langmuir 11* 3027-34
[5] Zangwill A 1996 *Physics at Surfaces* ed. Cambridge University Press Chapter 1 p 11.

6. Acknowledgements

This work was financially supported by CONACYT-Mexico under proyect 46224.

Synthesis of acrylic and allylic bifunctional cross-linking monomers derived from PET waste

A Cruz-Aguilar[1], A M Herrera-González[1], R A Vázquez-García[1], D Navarro-Rodríguez[2] and J Coreño[1,*]
[1] Universidad Autónoma del Estado de Hidalgo. Área Académica de Ciencias de la Tierra y Materiales. Carretera Pachuca – Tulancingo, Colonia Carboneras, Mineral de la Reforma, Hgo. México.
[2] Centro de Investigaciones en Química Aplicada. Blvd. Reyna Hermosillo. Saltillo, Coah. México.

[*]e-mail: jcorenoalonso@gmail.com

Abstract. An acrylic and two novel allylic monomers synthesized from bis (hydroxyethyl) terephthalate, BHET, are reported. This was obtained by glycolysis of post-consumer PET with boiling ethylene glycol. The bifunctional monomer bis(2-(acryloyloxy)ethyl) terephthalate was obtained from acryloyl chloride, while the allylic monomers 2-(((allyloxi)carbonyl)oxy) ethyl (2-hydroxyethyl) terephthalate and bis(2-(((allyloxi)carbonyl)oxy)ethyl) terephthalate, from allyl chloroformate. Cross-linking was studied in bulk polymerization using two different thermal initiators. Monomers were analyzed by means of ^{1}H NMR and the cross-linked polymers by infrared spectroscopy. Gel content higher than 90% was obtained for the acrylic monomer. In the case of the mixture of the allylic monomers, the cross-linked polymer was 80 % using BPO initiator, being this mixture 24 times less reactive than the acrylic monomer.

1. Introduction

Recycling of PET has been widely studied during the last decade due to the huge amounts of PET waste produced globally and because products made from recycled plastics can result in 50–60% energy saving as compared to making the same product from virgin resin [1]. Reusing or recycling of PET packaging can be carried out in different ways. PET containers can sometimes be directly reused as bottles or, on the contrary, its energy content can be recovered by means of incineration. On the other hand, physical recycling involves grinding and melting to obtain different products such as fibers. At this respect, an interesting new approach is the use of PET fibers obtained directly from cutting PET bottles to reinforce cement based materials. A last PET recycling method is their chemical reprocessing, which involves the total depolymerization of waste PET to the monomers, or partial depolymerization to oligomers and other chemical substances [2]. The monomers and oligomers thus obtained can be re-polymerized to produce again PET for packaging applications, or they can be reacted with different monomers to produce starting reactants for a range of different applications. Examples of this chemical reprocessing is the production of unsaturated polyester (UP) resins for polymer concretes and polymer mortars [3], the synthesis of acrylate/methacrylate-terminated oligoesters that could be used as radiation curable oligomers [2], or the formation of alkyd–amino resins for coating applications [4]. In this work we present the synthesis, cross-linking and characterization of an acrylic and two novel allylic monomers, obtained from glycolyzed PET through novel synthetic routes, which could be used as cross-linking agents for plastics in order to improve their thermal stability.

2. Experimental

1.1. Synthesis of BAOET

Bis(2-(acryloyloxy)ethyl) terephthalate (BAOET) was synthesized according to the literature [5]. A solution of acryloyl chloride (0.89 g, 9.84 mmol), BHET (1 g, 3.93 mmol) and tryethyl amine (0.99 g, 9.84 mmol) in an excess of acetone was placed in a 250 mL one necked round-bottomed flask, with constant stirring and under argon atmosphere during 24 h at room temperature (Figure 1). The reaction mixture was filtered with acetone. The remaining liquid was distilled under reduced pressure and then recrystallized from ethanol-water (2:5), filtered and dried in a vacuum oven at 50 °C to give 1.2 g (3.30 mmol, 83.96 % yield) of yellowish crystals (m.p. 82-86 °C). ^{1}H NMR (chloroform-d) δ (ppm): 8.15 (s, 4H; aromatic), 4.50 (t, 4H O-CH_2), 4.55 (t, 4H O-CH_2), 6.4 (t, 2H, C=CH), 5.95 (t, 2H, C=CH_2), 6.2 (t, 2H, C=CH_2).

1.2. Synthesis of AOCOEHET and BAOCOET

Allylic monomers were synthesized according to the literature [5]. A solution of allyl chloroformate (1.18 g, 9.84 mmol), BHET (1 g, 3.93 mmol) and pyridine (0.77 g, 9.84 mmol) in an excess of acetone was placed in a 250 mL two necked round-bottomed flask, with constant stirring, under argon atmosphere and using a dean-stark trap during 24 h at room temperature. The reaction mixture was filtered with acetone. The remaining liquid was distilled under reduced pressure, washed with water at 70 °C and dried in a vacuum oven at 50 °C to give 0.55 g (1.4 mmol, 36.11% yield) of a viscous ambar liquid, which is a mixture of the monofuntional and bifunctional monomers 2-(((allyloxi)carbonyl)oxy) ethyl (2-hydroxyethyl) terephthalate (AOCOEHET) and bis(2-(((allyloxi) carbonyl)oxy) ethyl)terephthalate (BAOCOET), respectively (Figure 1). The ratio of the mixture is 4:1, AOCOEHET to BAOCOET, according to the ^{1}H NMR spectrum. ^{1}H NMR (chloroform-d) δ (ppm): 8.15 (s, 3.73H; aromatic), 4.48 (t, 2.93H O-CH_2), 4.57 (t, 2.93H O-CH_2), 4.58 (m, 2.93H, O-CH_2), 6.06 (m, 1H, C=CH), 5.28 (m, 1.07H, C=CH_2), 5.42 (m, 1.07H, C=CH_2).

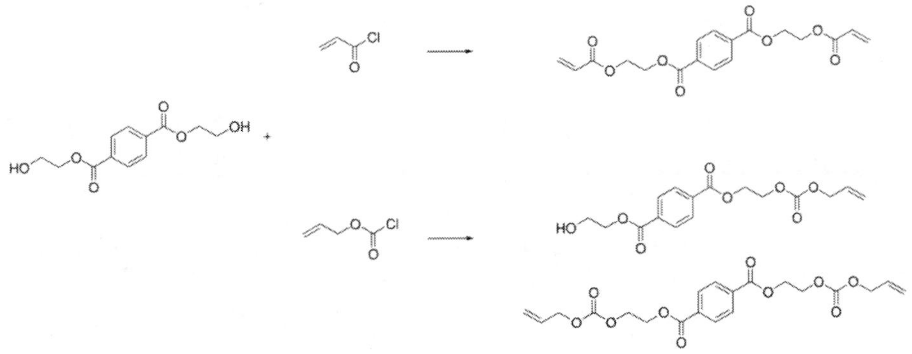

Figure 1. Reactions involving BHET with acryloyl chloride or allyl chloroformate producing BAOET or AOCOEHET/BAOCOET, respectively.

1.3. Bulk polymerization with thermal initiators

BAOET monomer was cross-linked in a quiescent polymerization. Square cells of 3 x 2.5 cm were constructed with two ordinary glass plates. Curable mixtures of BAOET were prepared by dissolving this monomer and two different initiatiors (benzoyl peroxide, BPO, and azobisisobutyronitrile, AIBN, 2 weight %) in acetone. The liquid mixtures were placed on glass plates, acetone was dried at room temperature, and the resulting solid mixtures were covered with a glass plate and clamped with ordinary paperclips. The monomers were polymerized by bulk thermal cross-linking into flat plates of 1-3 mm thickness at a reduced pressure of 42.7 kPa. A Stable Temp Cole-Palmer oven was used for thermal cross-linking at 90 °C.

The allylic monomers were polymerized in sealed Pyrex tubes using benzoyl peroxide (5 weight %) at 90 °C and AIBN (5 weight %) at 70 °C as initiators. Before polymerization, the tubes were degassed by the usual freezing and thawing technique under vaucum and then sealed off. The percentage of insoluble cross-linked polymer formed (gel content) was obtained by extracting the soluble part (sol fraction) from the cured monomer with acetone and filtration. After the extraction, the gel content was dried to a constant weight. Gel content is taken as the weight percent of polymer formed, % gel. Sol and gel fractions were characterized by IR. All the reaction systems were homogeneous before and after the polymerization.

3. Results and discussion

Figure 2a shows the ^{1}H NMR spectrum and the assignation of peaks for BAOET. Peaks D, E and F confirm the presence of the acrylic moiety, besides, it can be seen that the monomer is spectroscopycally pure through the integration of the signals corresponding to the 4 aromatic protons A and the 8 aliphatic protons C and B. In figure 2b it is observed the spectrum of the mixture of the allylic compounds. Peaks B, C, and D show a lower integration than expected for the bifunctional monomer thus peaks E, F and G show the presence of the allylic groups corresponding to the monofunctional and bifunctional monomers. The ratio AOCOEHET to BAOCOET (4:1) was calculated from the phenylene and allylic protons.

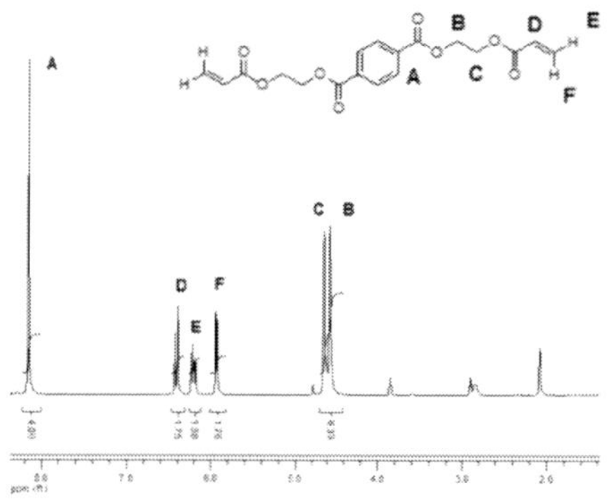

Figure 2a. ^{1}H-NMR spectrum for the BAOET monomer.

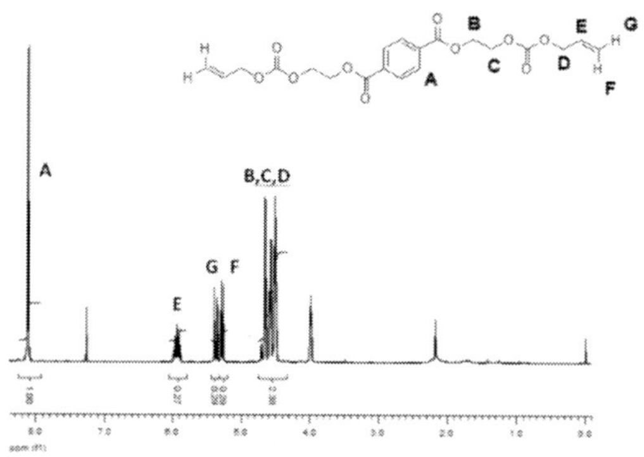

Figure 2b. ^{1}H-NMR spectrum for the AOCOEHET/BAOCOET monomers.

Cross-linked BAOET polymers were characterized by means of FT-IR spectroscopy (Figure 3) using samples with the same weight and also the same polymer:KBr ratio. After 20 min it is still possible to observe the bands at 700 and 3100 cm^{-1} corresponding to the C=C-H and C=C bonds respectively. The comparison of both spectra showed a noticeable intensity decrease, but not absence, of the alkene signals, even though the gel percentage approached 100%.

Figure 4 shows the variation of gel fraction as a function of time for the different monomers using BPO and AIBN. As can be seen, the mixtures of BAOET/BPO and BAOET/AIBN showed similar conversion times, over 90% before 20 minutes of polymerization. On the other hand, the mixture of allylic monomers/BPO reached a conversion nearby 80% after 6 h; when AIBN initiator was used, longer periods were required, obtaining a conversion of 2% after 6 h and, barely, 28% of gel content after 20 h (graph not shown Figure 4).

The noticeable differences between the percentages of gel formed can be due to the fact that the acrylic monomers are much more reactive than the allylic monomers. It is known that these monomers

IOP Conf. Series: Materials Science and Engineering **45** (2013) 012007 doi:10.1088/1757-899X/45/1/012007

show an internal inhibition or autoinhibition, which arises from the high resonance stability of the allylic radical formed during the polymerization [6]. It is expected that the cross-linked polymer obtained from the allylic monomer shows better mechanical properties and thermal stability than the crosslinked acrylic polymer due to the presence of the allylcarbonate moiety [5]. On the other hand, it seems that BPO is a better initiator for the allylic polymerization probably due to the fact that the half-life of AIBN is much lower than the half-life of BPO at the temperatures used during the polymerization [6]. Finally, it is possible that the presence of the monofunctional allylic monomer could also interfere in the common cross-linking behavior of the bifunctional monomer. Since the yield is still low for the allylic monomers, 13% for the bifunctional and 51% for the monofunctional, further work is being done in order to obtain a higher yield for the bifunctional monomer. The yield for the acrylic monomer was 83.6 %.

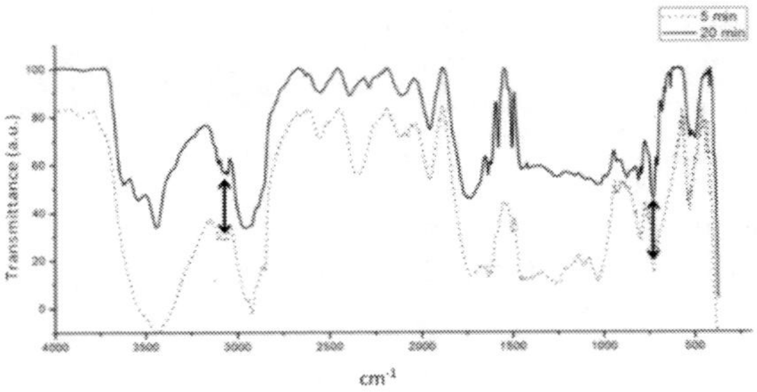

Figure 3. FT-IR spectra of BAOET polymer after 5 min (dot line) and 20 min (solid line). Arrows indicate the characteristic absorption bands of alkene groups bonds.

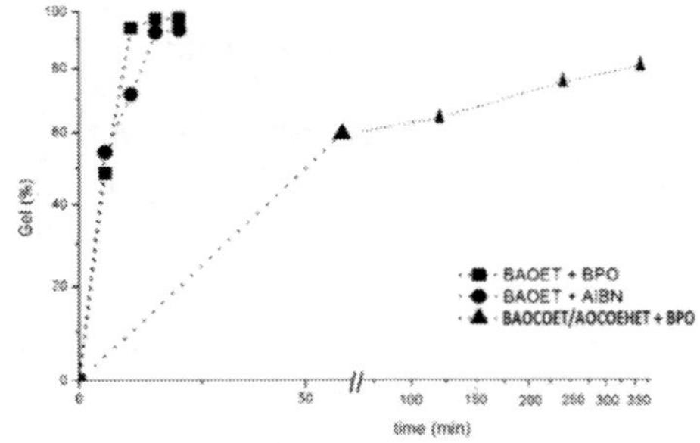

Figure 4. Cross-linking average of the mixtures of monomers with thermal initiators as a function of time: BAOET/BPO, BAOET/AIBN and BAOCOET/AOCOEHET/BPO

4. Conclusions

BHET obtained from PET depolymerization can be used to obtain bifunctional acrylic and allylic monomers, which could be used as cross-linking agents. In the case of the acrylic monomer, cross-linking rate and gel content is similar using both AIBN and BPO initiators, but for the allylic monomer cross-linking rate and gel content is faster using BPO than using AIBN. Cross-linking degree is higher for the acrylic monomer than for the mixture of allylic monomer, due to their different reactivities and to the presence of the monofuntional allylic monomer.

5. Acknowledgments

A. Cruz-Aguilar would like to express his sincere thanks to CONACyT for student grant.

6. References

[1] Karayannidis G, Achilias D 2007 *Macromol Mater Eng* **292** 128
[2] Nikles D, Farahat M 2005 *Macromol Mater Eng* **290** 13
[3] Abdelaal M, Sobahi T, Makki M 2011 *Construction and Building Materials* **25** 3267
[4] Torlakoğlu A, Güçlü G 2009 *Waste Management* **29** 350
[5] López-V D, Herrera-González A M, Castillo-Rojas S 2011 *Radiation Physics and Chemistry* **80** 481
[6] Odian G 1991 *Principles of polimerization* (New York: Wiley–Interscience) 266

IOP Conf. Series: Materials Science and Engineering **45** (2013) 012008 doi:10.1088/1757-899X/45/1/012008

Efficient method for polymerization of diallycarbonate and hexa(allylcarbonate) monomers and their thermal properties

A M Herrera-González[1]*, J García-Serrano[1], A A Pelaez-Cid[2] and I Montalvo-Sierra[2]

[1]Centro de Investigaciones en Materiales y Metalurgia, Universidad Autónoma del Estado de Hidalgo. Carretera Pachuca-Tulancingo Km. 4.5, 42184, Mineral de la Reforma, Hgo., México.
[2]Facultad de Ingeniería, BUAP, Puebla, México.
[3]Posgrado en Ciencia de Materiales, Universidad Autónoma del Estado de Hidalgo. Carretera Pachuca-Tulancingo Km. 4.5, 42184, Mineral de la Reforma, Hgo., México.

E-mail: mherrera@uaeh.edu.mx

Abstract

Polymers have had a significant increase in the industrial field, and polycarbonates (PCs) are an example. PCs have important properties such as high impact resistance, high heat capability, hardness, toughness, optical clarity, heat resistance, abrasion resistance and high refractive index. PCs with these properties have a variety of applications, for example, astronaut helmets, CDs/ DVDs, automobile parts, bulletproof windows, ophthalmic lenses and welding masks. In the present work we report the synthesis and characterization of network polycarbonates, based (1,1'-biphenyl)-4,4'-diyldiallylcarbonate (I) and hexa(4-allylcarbonatephenoxy) cyclotriphosphazene (III). The polycarbonates were obtained by bulk polymerization technique. The polymers obtained with thermal initiators showed degrees of cross-linking greater than 80%. The PCs obtained have thermal stability to 250 °C. The polycarbonate obtained from the monomer III showed a limiting oxygen index of 46.3%, therefore this polymer can be considered potentially useful as flame-retardant.

1. Introduction

Polycarbonates have been synthesized by different synthetic routes such as: reaction of diols with phosgene, copolymerization of epoxides with carbon dioxide in the presence of organometallic catalysts, ring opening polymerization of cyclic carbonate monomers, condensation of diols with CO_2, and radical polymerization of the vinyl or allyl carbonate monomers.[(1-3)] Monoallylic compounds do not form homopolymers of high molecular weight under free-radical initiation. However, the reactivity of the allylcarbonate monomers is enhanced with the increase of allylic groups, triallylcarbonate>diallylcarbonate >monoallylcarbonate. The radical polymerizations of allylcarbonate monomers have been studied using different initiators, such as peroxide, azo and peroxydicarbonate compounds, but due to their relatively low reactivity of this allyl compounds, high concentrations of initiators are needed (>3%) to be polymerized in a reasonable time. Other alternative for synthesis of PCs via free radical is the use of UV-Vis radiation and high energy radiation (gamma and electron radiation). In this work we are reporting the syntheses of allylcarbonate monomers, the syntheses of polycarbonates network, and thermal stability of network. The polycarbonates were obtained by bulk polymerization technique using benzoyl peroxide as initiator.

IOP Conf. Series: Materials Science and Engineering **45** (2013) 012008 doi:10.1088/1757-899X/45/1/012008

2. Experimental Section

2.1 Materials

All of the chemicals and solvents were reagent grade. Hydroquinone, hexachlorocyclotriphosphazene and 4,4′-biphenol were used as received without further purification. Allyl chloroformate and pyridine were distilled before use. Acetone was dried with potassium carbonate and distilled prior to use. Allyl(4-hydroxyphenyl) carbonate (II) was synthesized in the lab as described in the literature.

2.2 Instruments

Fourier transform infrared (FTIR) spectra were recorded on a Perkin Elmer FT-IR System 2000, using a KBr plate. ^{1}H NMR (400MHz), ^{13}C NMR (400MHz), and ^{31}P NMR (400MHz), spectra were obtained with a Varian Mercury spectrometer. As convention, the ^{1}H NMR and ^{13}C NMR spectra were referenced to tetramethylsilane, chloroform CDCl3 or acetone-d_6 as the solvent. Thermogravimetric study was carried out on a Mettler–Toledo (851) equipment with nitrogen as the purge gas at a 10 °C min-1 scanning rate from 25 to 800 °C. Elemental analysis was obtained with a Perkin Elmer analyzer (2499 C, H, N, Serie II), using acetanilide reference.

2.3 Synthesis of (1,1'-biphenyl)-4,4'-diyldiallylcarbonate monomer (I)

The compound I was synthesized by a Schotten-Baumann reaction, Figure 1a. A solution of 4,4-biphenol (1.5 g, 8.1 mmol), pyridine (1.29 mL, 6 mmol) and 40 mL of acetone, was placed in a 100 mL round-bottomed flask provided with stirrer and argon inlet. Then, allyl chloroformate (1.7 mL, 15.5 mmol) to the mixture was added. The reaction was maintained at 5 °C for 24 h under argon atmosphere. The solution was concentrated in vacuum yielding the solid diallyl compound I. The compound I was purified by column chromatography by SiO_2 using a mixture of dichloromethane-acetone (90:10). NMR ^{1}H ($CDCl_3$) δ (ppm): 4.78 (d, 2H, *J*= 1.2 Hz, CH_2=$CHCH_2$), 4.76 (d, 2H, *J*=1.2 Hz, CH_2=$CHCH_2$), 5.36 (d, $2H_{trans}$, J=17.2 CH_2=$CHCH_2$), 5.47 (d, $2H_{cis}$, *J*=10.4 Hz, CH_2=$CHCH_2$), 6.08-5.96 ppm (m, 2H, CH_2=$CHCH_2$), 7.26 (d, 4H, *J*= 7.2 Hz, aromatic), 7.55 (d, 4H, *J*= 7.2 Hz, aromatic). NMR ^{13}C ($CDCL_3$) δ (ppm): 69.25 (CH_2=$CHCH_2$), 119.65 (CH_2=$CHCH_2$), 121.38 (aromatic), 128.2 (aromatic), 131.02 (CH_2=$CHCH_2$), 138.22 (aromatic), 150.57 (aromatic), 153.47 (-C=O). FTIR (cm^{-1}): 3080 (υC-H), 2950 (υC-H), 1761 (υC=O), 1649 (υ C=C allyl), 1602 (υ C=C aromatic), 1272 (υ -C-O-C-), 933 (υ aromatic). m. p. 81-83 °C.

2.4 Synthesis of hexa(4-allylcarbonatephenoxy)cyclotriphosphazeno monomer (III)

The compound III was synthesized in two steps, Figure 1b. In the first step the compound II was prepared, as described in the literature.(4) In the next step, compound III was prepared following the next procedure. Compound II, 0.49 g (2.5 mmol), activated potassium carbonate 0.43 g (2.5 mmol) and 30 mL of THF, were placed in a three-necked flask provided with a stirrer, a reflux condenser, and an argon inlet. The mixture was maintained for 1h at reflux and stirred under argon atmosphere. After this time, hexa chlorotricyclotriphosphazene 0.14 g (0.4 mmol) was added. After reaction at reflux for 72 h, the mixture was cooled at room temperature, the potassium carbonate was separated by filtration, and the solvent was removed under distillation. The compound was purified by column chromatography by SiO_2 using a mixture of dichloromethane-acetone (80:20). Compound III was dried in vacuum at 60 °C, a color less liquid was obtained. Yield: 46%. RMN ^{1}H (Acetone-d_6) δ (ppm): 4.63-4.58 (m, 1H, CH_2=$CHCH_2$), 4.57-4.55 (m, ^{1}H, CH_2=$CHCH_2$), 5.33-5.24 (m, 1H, CH_2=$CHCH_2$), 5.20-5.10 (m, 1H, CH_2=$CHCH_2$), 5.95-5.83 (m, 1H, CH_2=$CHCH_2$), 6.95-6.87 (m, 2H, Aromatic), 6.78-6.60 (m, 2H, Aromatic). RMN ^{31}P (Acetone-d_6) δ (ppm): 7.04 (s, P=N). FT–IR (cm^{-1}): 3084 (υC-H), 2963 (υC-H), 1761 (υC=O), 1648 (υC=C allyl double bond), 1501 (υC=C Aromatic) 1243 (υ-C-O-C-), 1196, 1178 y 1165 (υP=N cyclotriphosphazene ring), 856 (υ Aromatic).

2.5 Synthesis of the 1,1'-biphenyl-4,4'-diallylcarbonate homopolymer (IV)

The compound II was polymerized in sealed pyrex tubes using benzoyl peroxide as initiator (5 weight %). The tubes were degassed by the usual freezing and thawing technique under vacuum and then sealed off. Tubes were taken out at different times and then the polymerized samples (cured monomer or castings) were analyzed. The percentage of insoluble crosslinked polycarbonate formed (gel content) was obtained by extracting the soluble part (sol fraction) from the cured monomer with CH_2Cl_2 and filtration. After the extraction, the gel content was dried to a constant weight. Gel content is taken as the weight percent of polymer formed, % gel. Sol and gel fractions were characterized by IR and elemental analysis. All the reaction systems were homogeneous before polymerization as well as at the end of polymerizations. FTIR (cm^{-1}): 3080 (υC-H), 2950 (υC-H), 1761 (υC=O), 1602 (υC=C aromatic), 1272 (υ-C-O-C-), 933 (υ aromatic).

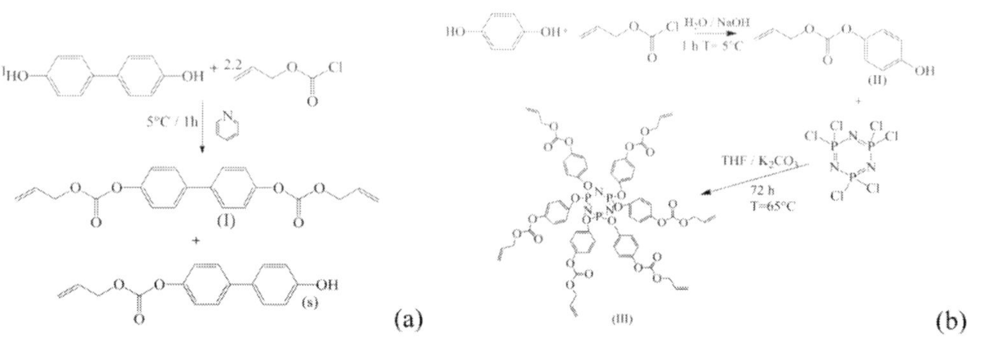

Figure 1. Synthetic route of (1,1'-biphenyl)-4,4'-diyldiallylcarbonate (a), and cyclomatrix phosphazene hexa(4-allylcarbonatephenoxy)ciclotriphosphazene (b)

2.6 Synthesis of the phosphazenecyclomatrix homopolymer (V)
100 mg of the compound III and 5 mg de benzoyl peroxide (5 weight %) were placed in two necked flask, equipped with a stirrer, thermometer and argon inlet. The temperature of the reaction mixture was increased to 90 °C and stirred for 72 h. The polymer was formed as a white solid. The solid obtained was separated by filtration. The polymer was washed in acetone in order to remove the residual monomer. FT–IR (cm^{-1}): 2960 (υ C-H), 1758 (υC=O), 1500 (υC=C Aromatic) 1262 (υ -C-O-C-), 1196, 1178 y 1165 (υP=N) 856 (υ cyclotriphosphazene ring).

3. Results and discussion

3.1 Monomers characterization
The compound (1,1'-biphenyl)-4,4'-diyldiallylcarbonate monomer is a white solid (90% yield). The hexa(4-allylcarbonatephenoxy)cyclotriphosphazeno monomer (III) is a colorless, viscous liquid. It was obtained with a purity of 99.8% and 60% yield. The chemical structures of monomers I and III were characterized by FTIR spectroscopy, ^{1}H, ^{13}C and ^{31}P NMR spectroscopy. Results are present in the experimental section.
Formation of allylcarbonate monomers is evident from their ^{1}H NMR spectra (Figure 2). The most important signals for both allylcarbonate monomers were the multiplets assigned to the protons of CH_2= (H_D) allylic group at 5.36 and 5.47 ppm for compound I, and at 5.33 and 5.20 ppm, for the monomer III, besides the signal of -CH- allyl group (H_C) was found in the range 6.08-5.96 ppm for the monomer I and in 5.95-5.83 ppm for the monomer III.
3.2 Polymers characterization
Bulk polymerization of monomers I and III using 5 weight % of benzoyl peroxide as initiator were studied at 90 °C. After polymerization, the cross-linked polycarbonates obtained from compounds I and II were practically colorless and insoluble in organic solvents. For the polymer IV the crosslinking was 96% after three hours of polymerization, whereas using reaction times lower than two hours the degree of crosslinking was negligible. The homopolymer V was obtained in 34 h, its formation was evident from typical FTIR. The absorption bands at 1205, 1176 and 1159 cm^{-1} were attributed to P=N and P-N bonds in phosphazene rings. The absorption band at 1260 cm^{-1} was attributed to the υC-O-C-esther group. Other evidence in FTIR was the absence of the signal at 1649 cm^{-1} corresponding to the υC=C of allyl group.

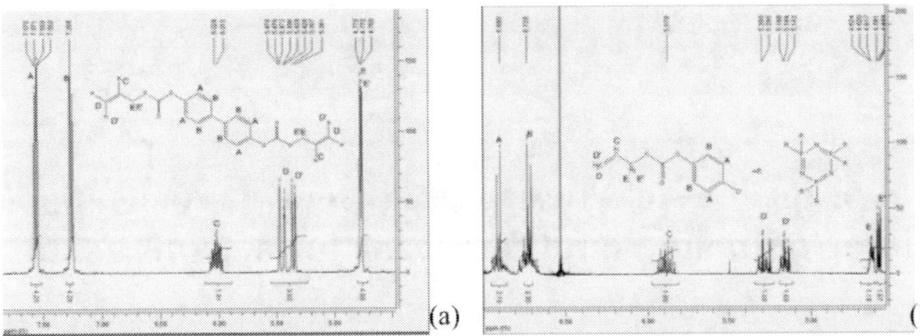

(a) (b)

Figure 2. Spectra ^{1}H NMR of monomers (1,1'-biphenyl)-4,4'-diyldiallylcarbonate (a), and cyclomatrix phosphazene hexa(4-allylcarbonatephenoxy)ciclotriphosphazene (b)

Thermal properties of the polymers obtained were analyzed by TGA. Figure 3 shows a comparison between the TGA curves of the polymers obtained. The polycarbonates exhibit good thermal stability as is reported in the literature.[5] The polymer IV is stable to 250 °C, while the polymer V begins to decompose at 240 °C. Also is interesting the behavior at 800 °C, polymer IV has lost 90% of its total weight, whereas the Polymer V has lost 28% of its weight. The result show that the polycarbonate with cyclomatrix has highest thermal stability, and the polycarbonate *V* could present the property of the flame retarder.[6] The limiting oxygen index (LOI) calculated via char yield data at 800 °C, using the semi-empirical formula proposed by Ven Krevelen,[7] was 46.3%.

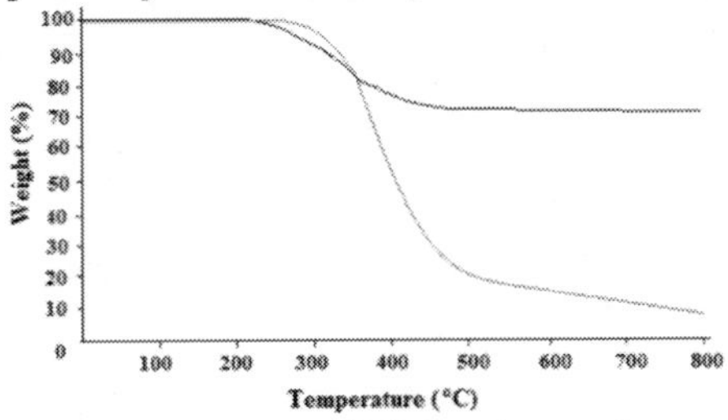

Figure 3. TGA of *1*,1'-biphenyl-4,4'-diallylcarbonate homopolymer (red) and phosphazenecyclomatrixhomopolymer (black)

4. Conclusions

(1,1'-biphenyl)-4,4'-diyldiallylcarbonate and cyclomatrixphosphazenehexa(4-allylcarbonatephenoxy) ciclotriphosphazene were easily synthesized in high yields and were bulk polymerized by thermal crosslinking with free radical initiators such as benzoyl peroxide. The PCs obtained have thermal stability to 250 °C. The polycarbonate V showed a limiting oxygen index of 46.3%, therefore this polymer can be considered potentially useful as flame-retardant.

Acknowledgments

The authors acknowledge financial support from **CONACyT** project **J010.195**. Montalvo-Sierra I would like to express her sincere thanks to CONACyT-32199 for Student grant.

References

[1] Cohen C T, Chu T and Coates G W 2005 *J. Am. Chem. Soc.* **127** 10869
[2] Darensburg D J 2007 *Chem. Rev.* **107** (6) 2388
[3] Boyles D A, Filipova T S, Bendler J T, Longbrake G and Reams J 2005 *Macromolecules* **38** (9) 3622
[4] Herrera-González A M 2006 Preparación de nuevos dialilcarbonatos y su polimerización por radicales libres *Tesis Doctoral* Puebla Pue. México
[5] Yonemori S, Masui A, Noshiro M 1991 J. Appl. *Polym. Sci. Appl. Polym. Symp.* **48** 523
[6] Matsumoto A, 2000 *Macromolecular Rapid Communications* **21** 1248
[7] Krevelen V 1990 *Properties of polymer: their correlation with chemical structure: their numerical estimation and prediction from additive group contribution* 3th Elsevier 73

IOP Conf. Series: Materials Science and Engineering **45** (2013) 012009 doi:10.1088/1757-899X/45/1/012009

Optical properties of a new p-phenylenevinylene oligomer containing quinolines and ferrocene for organic solar cells

F Zamora[1], R A Vázquez [1,*], J Coreño[1], I Moggio[2], E Arias[2], R Flores[2] and A de León[2]
[1]Centro de Investigaciones en Metalurgia y Materiales, Carretera Pachuca-Tulancingo Km 4.5 Colonia Carboneras. Mineral de la Reforma, Hidalgo, C.P. 42184. México
[2]Centro de Investigaciones en Química Aplicada. Blvd. Enrique Reyna Hermosillo No.140 C.P.25294 Saltillo, Coahuila México
E-mail: rosaangelesv@hotmail.com

Abstract. In the present work, we studied a conjugated phenylenevinylene oligomer containing quinolines and ferrocene (2FcQF1C) as end groups. The optical properties were analyzed by UV-vis and fluorescence spectroscopy. The energies of the Highest Occupied Molecular Orbital (HOMO) and Lowest Unoccupied Molecular Orbital (LUMO) were calculated by the oxidation/reduction potentials obtained by cyclic voltammetry in solution. A prototype of photovoltaic device was realized using a mixture 2:3 w/w of the oligomer (2FcQF1C) with the commercial [6,6]-phenyl-C_{61}-butyric acid methyl ester (PCBM) with a typical configuration of ITO/2FcQF1C:PCBM/Al. Current density-voltage (JV) curves were obtained in dark and using a laser at λ = 405 nm as illumination power source. Preliminary values of conversion efficiencies around of 10^{-4} % were obtained. These results are promising for optimization studies of the cells.

1. Introduction

Since the oil crisis in the early 1970's, there has been an increasing interest in the development of photovoltaic solar cells, in order to use the cleanest of all possible sources of renewable energy that is solar radiation. Nowadays, the technology of silicon devices still dominates the world market with a photon-to-current (PCE) conversion in the range of 24%, but its principal disadvantage is the high cost of production. As alternative materials, organic conjugated systems allow the fabrication of devices through less expensive techniques such as spin coating, doctor blade and ink jet printing. Despite that the PCE conversion of organic photovoltaic devices has passed from 0.001% in 1975 to 10.3 % in January 2012, it is still far from the values obtained with the inorganic counterparts. The origin of the lower PCE in organic photovoltaic cells is mainly due to the limited range of absorption of these materials. Several approaches have been proposed to overcome this problem such as combine different conjugated polymers in tandem solar cells [1], preparing composites or blends with quantum dots or metallic nanoparticles, and/or lowering the optical band gap by inserting electron donor groups in the conjugated chain. Ferrocene is a good electron donor and photoinduced electron transfer has been demonstrated for ferrocene bearing materials. Recently, charge transfer in ferrocene substituted poly(thiophene)s have been demonstrated in bilayer solar cells. Also, improved open circuit voltages have been obtained in heterojunction photovoltaic devices made of ferrocene as electron donor and fullerene as electron withdrawing material.

In this contribution we show that ferrocene strongly affects the optical and electrochemical properties of a novel phenylenevinylene oligomer with respect to its precursor that does not present in its structure ferrocene end groups. Particularly, the absorption spectrum becomes broad and unresolved, the fluorescence quantum yield decreases as well as the band gap. The free energy for intermolecular photoinduced energy transfer with PCBM as calculated by the Weller equation indicates that only the heptamer is a possible candidate for solar cells application. The intrinsic photovoltaic properties are studied in a heterojunction prototype.

2. Experimental

The chemical structure of the oligomer 2FcQF1C is reported in the figure 1b together with its precursor pentamer 2BrQF1C (figure 1a) that was studied as a term of comparison for optical and electrochemical characterization.

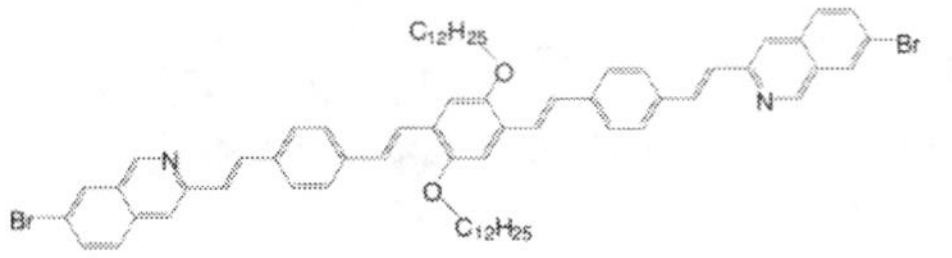

2BrQF1C

Figure 1a. Structure of 2BrQF1C pentamer.

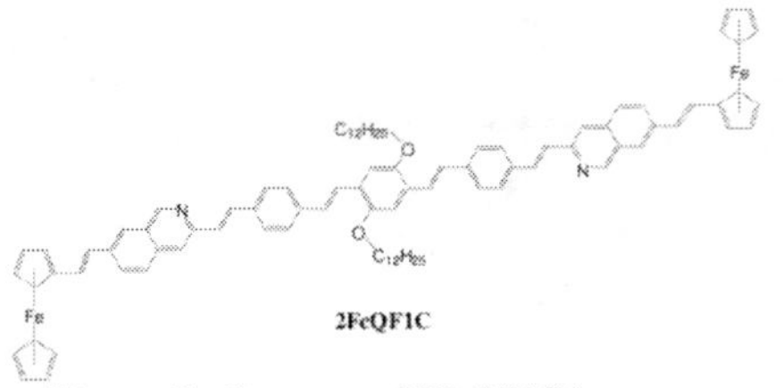

2FcQF1C

Figure 1b. Structure of 2FcQF1C heptamer.

2.1 Optical characterization

The absorption and emission spectra were obtained in $CHCl_3$ and thin films using a Shimadzu 2401PC spectrophotometer and a Perkin Elmer LS50B spectrofluorimeter, respectively. For the fluorescence spectra, the samples were excited at 10 nm below the absorption maxima. Fluorescence quantum yield in solution was obtained according to the literature method [2]. The temperature was controlled to 25.0±0.3 °C with a water circulating bath. The optical band gap $E_{g,op}$ was obtained from the intersection between the X axis and the tangent at the normalized absorption curve drawn in correspondence of optical density of 0.1.

2.2. Electrochemical characterization

The electrochemical properties of oligomers were investigated at room temperature in 0.5 mM CH_2Cl_2 solutions by cyclic voltammetry in a C3 Stand cell from Basi coupled to a potentiostat/galvanostat ACM Gill AC. Bu_4NPF_6 (0.1M) was used as the supporting electrolyte, a glassy carbon as working electrode, Pt wire as counter electrode, Ag/AgCl as the reference electrode and ferrocene/ferrocenium as internal reference. The experiments were carried out in a nitrogen atmosphere at a scanning rate of 50 mV/s and electrochemical window from 2500 mV to -1500mV beginning with a reduction process. The molecular orbital energies HOMO and LUMO were calculated from the oxidation (Eox) and reduction (Ered) potentials with the equation 1:

$$E^{HOMO[LUMO]} = -e\left((E^{ox[red]}_{1/2(vs\ Ag/AgCl)} - E^{ox}_{onset(Fc/Fc^{+}\ vs\ Ag/AgCl)}\right) - 4.8eV \quad 1$$

The free energy (ΔG)associated with the possibility to have intermolecular photoinduced electron transfer has been calculated with the electrochemical and fluorescence information modifying the Weller equation (equation 2) by considering that, for intermolecular process, the distance anion-cation trends to infinite [3]:

$$\Delta G = -e(E^{D}_{ox} - E^{A}_{red}) - E_{10} - \frac{e^2}{8\pi\varepsilon_0}\left(\frac{1}{r^-} + \frac{1}{r^+}\right)\left(\frac{1}{\varepsilon_r} - \frac{1}{\varepsilon_s}\right) \quad 2$$

The cation (electron donor material) radium, r^+ has been obtained by simulating the equilibrium geometry of the molecules at AM1 semiempirical level in Spartan 10. The anion (electron withdrawing material) is in our case PCBM and its radium has been reported to be 5.6Å [3].

2.3 Solar Cells
Devices were prepared with the configuration ITO/PEDOT:PSS/active layer/Al. PEDOT:PSS (5:8) with a $\sigma = 1$S/cm and dispersed in water (1.3% weight) from Aldrich was deposited on the Indium Tin Oxide (ITO) by spin coating at 1000 rpm to give a 25 nm layer. Active layers with thickness around 160 nm were obtained from a 2:3 weight mixture of 2FcQF1C and commercial PCBM (Aldrich) dissolved in 1:1 toluene/ $CHCl_3$ mixture. The aluminum was vacuum evaporated at a rate of 3 Å/s at a typical pressure of 10^{-6} torr in an Intercovamex TE18P vacuum chamber; the thickness was controlled by the quartz balance monitor ~100 nm. The current density-voltage (J-V) curves were obtained using a Keithley 6517A electrometer by illuminating the devices from the ITO side with 405 nm Newport laser (40 mW power).

3. Results
3.1 Optical characterization in solution
2BrQF1C can be considered, in terms of its optical properties, as a sum of two chromophores: the quinoline-phenylenevinylene conjugated part and ferrocene. The first one is also a fluorophore while ferrocene does not emit. The UV-Vis spectrum of 2BrQF1C (figure 2, dashed line) shows an excitonic spectrum with a maximum at 430 nm, similarly to other phenylenevinylene pentamers. On the contrary, the absorption spectrum of 2FcQF1C presents a broad unresolved absorption, suggesting electronic interactions in the ground state between the two chromophores. The optical band gap $E_{g,op}$ decreases from 2.41 eV to 2.36 eV and the fluorescence maximum (inserted figure) red shifts from 507nm to 517 nm, in agreement with the increase in the conjugation length. The incorporation of ferrocene affects the fluorescence quantum yield that decreases from 0.31 (2BrQF1C) to 0.03 (2FcQF1C). A similar behaviour was previously found in other conjugated systems and ascribed to photoinduced energy transfer [4].

3.2 Electrochemical characterization
The cyclic voltammetry studies allowed us to calculate the values of reduction and oxidation potential for the determination of the electrochemical band gap. Table 1 shows the electrochemical properties determined from the voltagrams.

Table 1. Electrochemical properties of 2BrQF1C y 2FcQF1C in CH_2Cl_2.

Compound	E^{ox}_{max} (V)	E^{red}_{max} (V)	E^{ox}_{onset} (V)	E^{red}_{onset} (V)	$E^{ox}_{1/2}$ (V)	$E^{red}_{1/2}$ (V)	HOMO (eV)	LUMO (eV)
2BrQF1C	2.30	-1.10	1.82	-0.87	2.06	-0.98	-6,86	-3,31
2FcQF1C	0.85	-0.66	0.64	-0.45	0.75	-0.55	-5,55	-4.25

2BrQF1C presents one oxidation/reduction peak associated with the removal/addition of one electron to the conjugated chain giving a polaron excited state. 2FcQF1C shows two oxidations peaks (maxima at 0.85 and 1.25 V) that can be ascribed to polaron and bipolaron formation. The first oxidation peak is shifted with respect to the ferrocene as found for other ferrocene substituted molecules. The semiempirical molecular simulation gave a cation radius of 16.5Å and 20.4 Å for 2BrQF1C and 2FcQF1C respectively. These values were introduced in equation 3 together with the oxidation and reduction potentials obtained by voltammetry and the value for the E_{10} (2.57 eV for 2BrQF1C and 2.67 eV for 2FcQF1C) obtained from the optical study, to give the free energy $\Delta G = 0.36$eV for 2BrQF1C and -1.21eV for 2FcQF1C. As photoinduced energy transfer is spontaneous when ΔG is negative, this result suggests that only 2FcQF1C is a good candidate for solar cells application.

3.3 Cells
On the basis of the previous discussions, solar cells were prepared only for 2FcQF1C in heterojunction with PCBM. It is to be noted that the prototypes were fabricated in laboratory conditions, without

optimization of the film preparation just in order to study the intrinsic optoelectronic properties of the new molecule. The active layer presents an increase in the absorption in the 450-800 nm with respect to the film of 2FcQF1C (figure 3), which permits an antenna effect in the solar maximum emission (that is around 500 nm).

The solar cell prototype shows the typical J-V curve (figure 4a) with a low V_{OC} (0.67 V). However the saturated current density (3.92 x 10^{-3} mA/cm^2) and the efficiency (10^{-4} %) are lower than usually reported for OPVs. This can be due to microsegregation phenomenon with PCBM as found by Atomic Force Microscopy (AFM, figure 4b). This is to be pointed out that the present prototype was not optimized in terms of the film preparation conditions. Morphology has been demonstrated to be an important factor to reduce the exciton diffusion inside the bulk heterojunction. Changing the solvent, using additives or annealing can improve this aspect and therefore the efficiency [5]. For this reason, works in progress are addressed to investigate the better conditions for the film preparation (deposition technique, solvent, annealing, etc) for increasing the device efficiency.

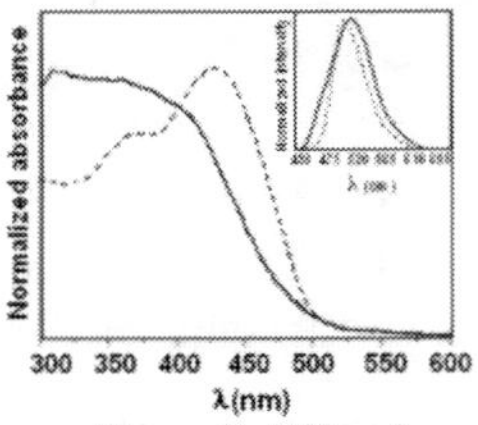

Figure 2. UV and fluorescence (inserted) spectra of 2FcQF1C (solid lines) and 2BrQF1C (dashed lines).

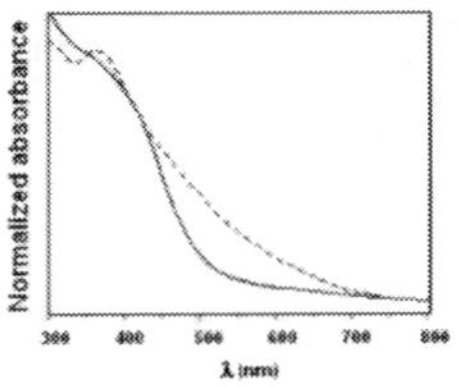

Figure 3. Absorption spectra of 2FcQF1C (solid line) and 2FcQF1C:PCBM 2:3 (dashed line) film.

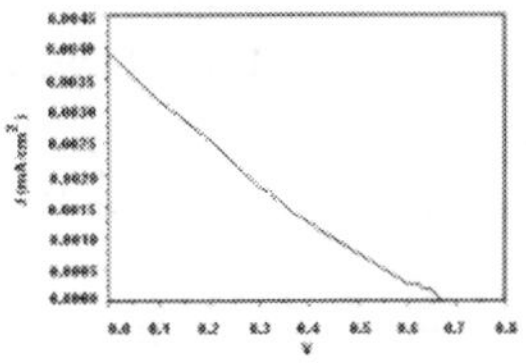

Figure 4a. J-V curve of a 2:3 2FcQF1C:PCBM prototype solar cell.

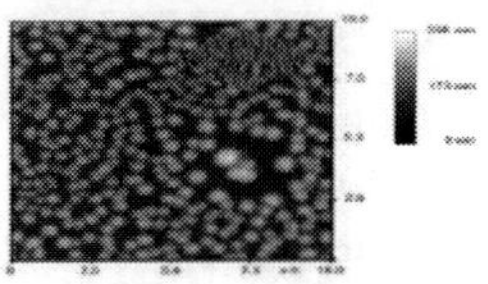

Figure 4b. Bi- and tri(inset)dimensional AFM image of the active layer.

3.3. Conclusions

The optical and electrochemical properties of a conjugated phenylenevinylene heptamer bearing ferrocene and quinoline were studied and compared with its precursor pentamer, which is not substituted with ferrocene. The free energy for intermolecular photoinduced energy transfer to PCBM was calculated for both molecules by inserting the oxidation/reduction potentials and the first excited state energy obtained by fluorescence in the Weller equation. According to this study, just the ferrocene heptamer is a good candidate for solar cells application. The efficiency of the prototype solar cell based on heterojunction with commercial PCBM was low (around 10^{-4} %), most likely due to microsegregation process between the oligomer and the fullerene derivative. Optimization of film preparation conditions is currently in progress.

4. References

[1] Wu MH, Ueda A and Mu R 2005 Semiconductor Quantum Dots Based Nanocomposite Solar Cell *Organic Photovoltaics Mechanisms, Materials and Devices*, eds SS Sun and NS Saricifitci (Boca Raton: Taylor and Francis Group) chapter 14

[2] Peeters E, van Hal O-A, Knol J, Brabec CJ, Sariciftci NS, Hummelen JC and Janssen R 2000 *J. Phys. Chem. B* **104** 10174

[3] Jiu T, Li Y, Liu X, Liu H, Li C, Ye J and Zhu D 2007 *J. Polym. Sci. A: Polym. Chem.* **45** 911

[4] Chandrasekhar P 1999 *Conducting polymers, fundamentals and applications:A practical approach* (Boston: Kluwer Academic Publishers)

[5] Park SH, Roy A, Beaupre S, Cho S, Coates N, Moon JS, Moses D, Leclerc M, Lee K and Heeger A J 2009 *Nat. Photon.* **3** 297

Acknowledgement

Financial support from Conacyt CB84586 and 98513-R projects.

Third Congress on Materials Science and Engineering (CNCIM-Mexico 2012) IOP Publishing
IOP Conf. Series: Materials Science and Engineering **45** (2013) 012010 doi:10.1088/1757-899X/45/1/012010

Study Anisotropy of a ductile iron class 80-60-03 hardening

N Lopez Perrusquia[1], M A Doñu Ruiz[1], D Sánchez Huitron[2], J D Villegas Cárdenas[2], A Camarillo Villegas[2]

[1]Instituto Politécnico Nacional, SEPI-ESIME, U. P. Adolfo López Mateos, Zacatenco, México D.F., 07738, México.
[2]Universidad Politécnica del Valle de México, Grupo Ciencia e Ingeniería de Materiales, 54910, Tultitlán, Estado de México.

E-mail: [1]nlopezp0803@ipn.mx,[1]mdonur0800@ipn.mx

Abstract: This paper studies the effect of anisotropy in the propagation of surface cracks to identify areas of potential failure, and thus implement methods that increase the resistance of the phases created in the surface hardened cast iron ductile boride by the process. On the other hand assesses the influence of residual stresses using the fracture toughness microindentación by Palmqvist cracks regime taking the perpendicular and parallel cracks generated in the hard surface, with samples of 6 h and 8 h, 900 ° C treatment of boriding, with loads of 1.9 and 2.9 N. Where the value of at $(k_c)\left(\frac{\pi}{2}\right) > (k_c) > (k_c)(\theta)$ a constant distance of 30 microns. Also the phases present were determined by X-ray diffraction (XRD) and scanning electron microscopy (EDS), finally adherence is determined by the indentation technique HRC.

1.Introduction

The anisotropy and residual stresses in mechanical components of nodular iron castings, originate from various causes such as forming processes, manufacturing and fabrication, welding, machining, casting, heat treatment, cooling and heating irregular, among others decrease the ability loading of the metal pieces [1-2]. Furthermore various investigations have been performed to estimate nondestructive anisotropy and the influence of residual stresses in surface-hardened materials, using the technique using the indentation induced fracture under the regime Palmqvist type, however the evaluation of the cracks parallel and perpendicular generated by microindentation method determines the hardened surface anisotropy in materials used in manufacturing [3-4]. At the same time, it is known that the fracture surface of the treated materials under varying loads of indentation produced cracks in different directions perpendicular and parallel to the surface of the material. Therefore the evaluation of anisotropy in surface hardened metal materials can be obtained by employing the technique Vickers, using the criterion of the fracture toughness (KC) [5-6]. So it is important to control or decrease in the concentrations of residual stresses, in order to study the influence of anisotropy on surface hardened nodular iron by a nondestructive technique and low cost.

2. Experimental

2.1. Surface Hardening Treatment.

Specimens were prepared of a ductile iron Class A 536 Class 80-60-03 With dimensions of 12.7 mm diameter by 50.0 mm long for the treatment of hardening of boride. The treatment was performed in a muffle without conventional inert atmosphere, with temperature 900 ° C, with time 6 and 8h.

2.2. Fracture toughness

The study of anisotropy Fe_2B layer on the surface of cast iron class 536 80-60-03 A; was evaluated by employing the technique Vickers, using the criterion of the fracture toughness (K_C) [7]. Used models Palmqvist type system for this work, in table 1 presents the models used to evaluate Kc. In order to verify the presence of phase Fe2B iron on the surface of hardened, an analysis was made of Energy Dispersing Spectroscopy (EDS).

Table 1. Models to evaluate Kc

Models	References
$K_C = 0.015\left(\frac{g}{a}\right)^{1/2}\left(\frac{E}{H_V}\right)^{2/3}\left(\frac{P}{c^{3/2}}\right)$	[8]
$K_C = 0.0264(H_v a)\left(\frac{E}{H_V}\right)^{2/5} g^{1/2}$	[9]

2.4. Study of Adhesion Layer.

Evaluation of adhesion was performed with the technique Rockwell C. The results of the fractured layer is compared with the standard defined quality of adhesion, where HF 1 to HF 4 defines a sufficient adhesion, while the HF 6 and HF 5 show insufficient adhesion (HF is the abbreviation of the German force adherence).

3. Discussion and Results.

In figure 1 shows micrographs of hardened nodular iron, revealing the layer Fe_2B and morphology, this shape type of the layer must be high content of alloying elements in this ductile iron.

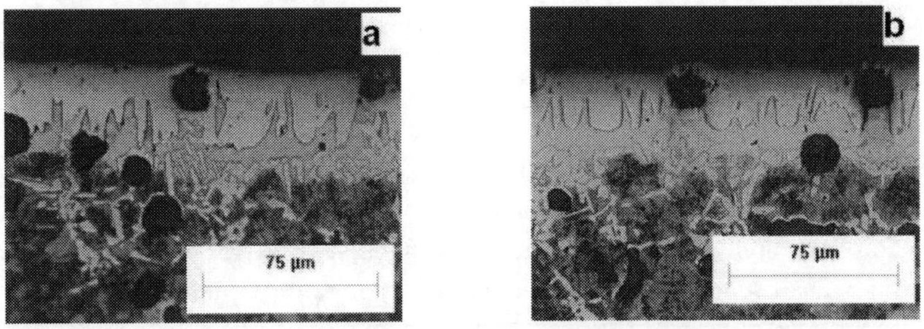

Figure 1. Micrographs of ductile cast iron superficially hardened by the process of boriding to 900 ° C, with times to (a) 8h and (b) 6h.

Table 2 presents the results of fracture toughness for the two models used for this study. Factor values for the stress intensity iron boride Fe2B to $\frac{\pi}{2}$ radians and θ obtained by the equation; K. Niihara and Laugier M.T.; With 6 and 8 hours of treatment at a distance from the surface of 30 microns.

Table 2. Values of fracture toughness perpendicular and parallel

Temperature	**Time**	**Load**	**K. Niihara (MPa m$^{1/2}$)**		**Laugier M.T. (MPa m$^{1/2}$)**	
(°C)	**(h)**	**(N)**	$\frac{\pi}{2}$	θ	$\frac{\pi}{2}$	θ
900	6	1.9	3.70±1.12	3.89±1.09	5.38±0.81	4.19±1.07
900	8	1.9	3.56±1.18	3.39±1.21	4.31±0.90	4.00±1.35
900	6	2.9	3.87±1.23	3.74±1.39	5.05±0.21	4.19±1.07
900	8	2.9	3.96±1.09	3.79±2.02	4.31±0.90	4.11±1.15

Moreover the percentage of the anisotropy is obtained in Table 3. To verify the dependence of the anisotropy with resistance to cracking Fe_2B borided layer, it was necessary to set the value of stress intensity factor critical of the form (KC) (π / 2) > (KC) > (KC) (θ), being π / 2 and θ the angle between the direction of crack propagation and the surface [11]. Depending on the shape coefficient of the grains of boride ductile cast iron, the degree of anisotropy of the resistance to cracking is determined through equation $\frac{\left(k_c\left(\frac{\pi}{2}\right) - k_c(\theta)\right)}{k_c(\theta)} = 30 - 80\ \%$ [11-12].

Table 3. Evaluation percentage of Anisotropy.

Temperature	**TIME**	**Load**	**Average values of K_C (MPA m$^{1/2}$)**		**Degree anisotropy**
(°C)	**(h)**	**(N)**	$\frac{\pi}{2}$	θ	**(%)**
900	6	1.9	4.68±1.05	4.17±1.38	12.2
900	8	1.9	3.98±1.14	3.88±1.17	2.5
900	6	2.9	4.56±0.82	4.05±1.23	12.5
900	8	2.9	4.28±1.04	4.05±1.38	5.6

EDS analysis obtained by SEM shown in figure 2. The results show that the alloying elements that are in phase with the formation Fe2B, carbon and silicon do not dissolve significantly during the phase and does not diffuse through the boride layer, being displaced to the diffusion zone, and forms together with boron, and solid solutions as silicoborides and boron cementite.

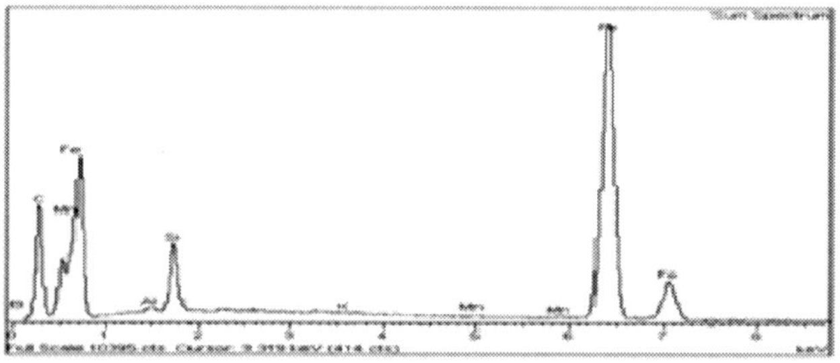

Figure 2. EDS spectrum of surface-hardened nodular iron

The technique is performed by the indentation test Rockwell-C [10], The indentations generated by this technique compressive stress and tension in the impact surface resulting in cracking, peeling, adhesion and different morphologies which studying this ductile cast iron, figure 3 has a sufficient adhesion HF 3.

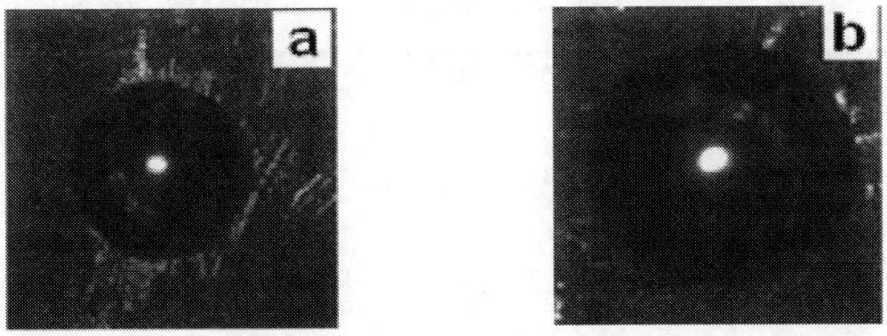

Figure 3. (a, b) Rockwell C indentations, evaluation of the adhesion of surface-hardened of cast iron ductile.

4. Conclusions

The values presented in the fracture toughness are slightly larger in the direction perpendicular to the surface where the cracking occurs at the vertices of the indentation. This indicates that at a constant distance indentation, there is no influence of residual stresses in the crack resistance behavior, remaining Kc invariant in relation to residual stress. Furthermore, the degree of anisotropy remains below 30%. It is therefore important to the estimated residual stress experimentally along the layer thickness borurada to determine quantitatively the influence of the efforts in the crack growth of the total Fe_2B phase. Besides the characteristics of the formation of cracks obtained are presented by the anisotropy of the layer formed on the surface, likewise is a reflection of its non-uniform composition, and the crystallographic and morphological anisotropy. Furthermore, the anisotropy and fracture toughness may be related residual stresses.

5. Acknowledgements

CONACyT and COMECYT the State of Mexico for the support of confinement to attend this Congress; also Promep.

6. References

[1] A.K. Sinha, 1991 Boriding (Boronizing), ASM Handbook, OH, USA. Journal of Heat. Treatment **4,** 437-447
[2] C Fischer R Schaaber 1976 Proceedings of Heat Treatment '76. The Metals Society 27 – 30
[3] A. V. Byakova et. al, 1989 Transla. Poroshk. Metall. **4** 58-63
[4] ASTM Standard Test Method for Microindentation Hardness of Materials (E 384 – 99)
[5] D. Golanski et. al. 1995 Jour. of Mater. Scien. Letters **14** 1499-1501
[6] N López-Perrusquia et al. 2009 Advan. Mater. Rese. **65** 47-52
[7] A G Evans and E A Charles 1976 Jour. of the Amer. Ceram. Soci. **59** 371-372
[8] Laugier M T 1987 Jour. of Mater. Scien. Letters, **63** 55-356
[9] K. Niihara 1983 Jour. of Mater. Scien. Letters **2** 221-223
[10] N Vidakis A Antoniadis N Bilalis 2003. Jour. of Mater. Proce. Techn. **143–144** 481–485
[11] A V Byakova 1993 Transla. Poroshk. Metall. **4** 36-46
[12] A V Byakova and G Gorbach Trans. Prob. Proch. **26** 51-61

IOP Conf. Series: Materials Science and Engineering **45** (2013) 012011 doi:10.1088/1757-899X/45/1/012011

Effect of annealing on structural properties of SiO_x films obtained by HWCVD technique

A Coyopol[1], T Díaz, G García, R López, H Juárez, E Rosendo, M Pacio, and G Escalante
CIDS-ICUAP, BUAP, Ciudad Universitaria, Ed. 103 D, Col. San Manuel, C.P. 72570, Puebla, México. Tel: (222) 229-5500 ext. 7855

[1]E-mail: acoyopol@gmail.com

Abstract: SiO_x films were deposited in the temperature range of 800-1000 °C by hot wire chemical vapor deposition (HWCVD) technique. Annealing treatment for as-deposited samples was performed in nitrogen (N_2) atmosphere at 1100 °C. The effect of annealing atmosphere on the structural properties of the deposited films was investigated by means of micro Raman, X-ray diffraction (XRD), Photoluminescence (PL) and high resolution electron microscopy (HRTEM) measurements. Micro-Raman studies revealed an increase in the crystalline volume fraction (X_c) in the deposited films after annealing. XRD patterns of the annealed films showed the formation of silicon crystals. HRTEM images confirmed the presence of silicon nanocrystals (Si-ncs) in both as-deposited and annealed films. Furthermore, the annealing treatment on the as-deposited samples at 900 and 1000 °C resulted in the observation a large number of Si-ncs with sizes between 3-15 nm. PL measurement shows a broad emission from 400 to 1100 nm. This emission was related with both Si-ncs and interfacial defects present in SiO_x films.

1. Introduction

Since the discovery of the luminescent properties of Si/SiO_2 multilayers [1, 2], Si-ncs embedded in SiO_x films have been extensively studied, mainly because they exhibit more stable optical properties than porous silicon. Recent publications [3, 4, 5], have suggested that SiO_x films with embedded Si-ncs can be an alternative in the development of silicon-based photovoltaic devices. Nanocrystalline material is more stable and has broader absorption range in the visible part of the solar radiation [5, 6]. In this direction, one of the most promising routes is exploiting the quantum confinement effect (QCE) in silicon nanocrystals. Some interesting effects based on the QCE are the multiple exciton generation in Si quantum dots (QDs) [4] or the optical band gap modulation with the QDs size [7].

Several techniques have been used in the production of Si-ncs embedded in SiO_x matrix such as: sputtering [8], ion implantation [2], plasma enhanced chemical vapor deposition (PECVD) [9], and HWCVD [10, 11].

In this work evolution of the structural properties of SiO_x films obtained by HWCVD were investigated. To achieve this goal, samples of SiO_x were annealed in N_2 atmosphere at 1100 °C. It was found that both the mean size and the number of Si-ncs is increased by effect of annealing process. These features should be considered to improve the quantum confinement effects and to absorb sunlight in a more efficient way in a solar cell.

2. Experimental Details

The HWCVD process has been already described elsewhere [11]. The filament-source distance was kept constant at 3 mm, while the filament-substrate distances were 1.32, 1.16 and 1.0 cm, to obtain deposition temperatures of 800, 900, and 1000 °C respectively. The growth time was 10 minutes on each experiment. The quartz substrates were carefully cleaned with a MOS standard cleaning process before being introduced into the reactor. After deposition the samples were annealed in N_2 atmosphere for two hours at a temperature of 1100 °C.

The XRD diffractograms were measured with a Bruker D8 Discover diffractometer using Cu Kα radiation (λ=0.15418 nm). Micro-Raman measurements were performed at room temperature using a He-Ne laser (632.8 nm). A laser with a wavelength of 405 nm and 40 mW of power was used to excite the sample in PL measurements; the range detected by monochromator was from 400 to 1100 nm. Finally, a high-resolution transmission electron microscopy (HRTEM) FEI Tecnai F30 STWIN G2 was used to observe the presence and evolution of the Si-ncs.

3. Results and Discussion

In our previous work structural and optical properties of Si-ncs in SiO_x films were reported [11], micro-Raman studies suggest that during the HWCVD process some formed silicon clusters can be crystallized to form Si-ncs due to an apparent in-situ annealing at growth temperatures of 900 and 1000 °C. In order to have a better understanding of this phenomenon, some samples were annealed at 1100 °C in N_2 atmosphere for two hours. Figure 1 shows micro-Raman spectra of samples grown at 800, 900, and 1000 °C with subsequent annealing at 1100 °C. The peak at 521 cm^{-1} related to the nanocrystalline phase is detected in the samples. It is noteworthy that before annealing, the samples growth at 900 and 1000 °C causes crystallization in silicon clusters [11]. When the samples are annealed both crystallization in silicon clusters and the number of Si-ncs increases. One factor that may give information about the previous discussion is the crystalline volume fraction (X_c) parameter.

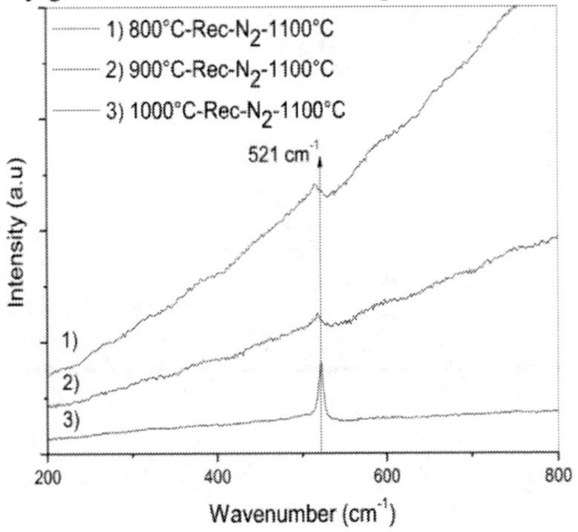

Figure 1. Micro-Raman spectrum of films growth at 800, 900 and 1000 °C with subsequent annealing at 1100 °C.

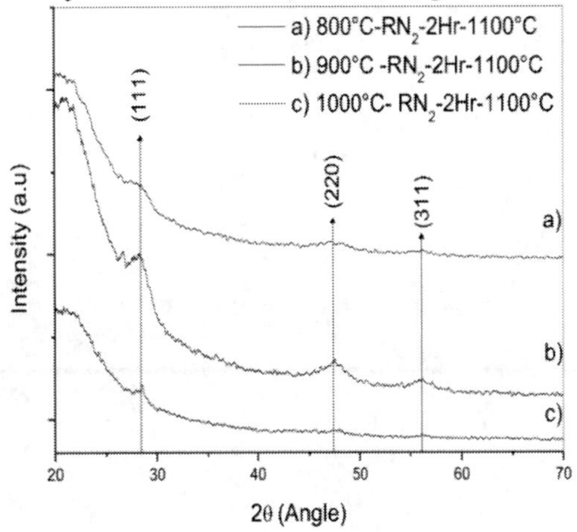

Figure 2. XRD patterns for samples deposited at 800, 900 and 1000 °C with subsequent annealing at 1100 °C.

The crystalline volume fraction X_c can be estimated from micro-Raman spectra. X_c is given by, $X_c=I_{TO2}/(I_{TO2}+I_{TO1}*m)$, where I_{TO2} is the integrated micro-Raman intensities of the crystalline peak, I_{TO1} the integrated micro-Raman intensities of the amorphous peak, m is a factor, which can be obtained from; m=0.1+exp(-d/25), where d (nm) is the diameter of the grain obtained from the XRD spectrum, and m=1 [6]. Table 1, shows the crystalline fraction for different conditions. For as-deposited samples the X_c factor increases from 4 to 10%. For samples annealed at 1100 °C, the X_c

factor increases from 27 to 50 % by changing the growth temperature. The increase in the X_c factor can be understood as a modification in the number of the Si-ncs as well as their mean size.

The mean size of the Si-ncs was investigated by XRD and HRTEM analyses. XRD characterization revealed the presence of crystalline planes in SiO_x films. Figure 2 shows XRD patterns for samples deposited at 800, 900 and 1000 °C with subsequent annealing at 1100 °C. Three main peaks are localized at 28.4, 47.4, and 56.1° corresponding to the silicon (111), (220), and (311) planes, respectively. Table 1, shows the data of the mean size of the Si-ncs obtained by Scherrer approximation for samples deposited at 800, 900, and 1000 °C with subsequent annealing at 1100 °C. As can be seen, the size of Si-ncs increase with the growth temperature. This is expected because in as-deposited films, the Si-ncs diameter is larger at 1000 °C than those growth at 900 °C [11]. Therefore, when the films are annealed, Si-ncs continues growing.

Table 1. Estimation of X_c and average size of the Si-ncs.

Temp. °C	As Deposited X_c (%)	Annealed at 1100 °C X_c (%)	Annealed (1100 °C) and Average Size of Si-ncs by XRD (nm)
800	----	27	3.24
900	4	30	4.39
1000	10	50	19.8

Figure 3 and 4 show the HRTEM images from samples deposited at 900 and 1000 °C with subsequent annealing at 1100 °C. It is confirmed the existence of Si-ncs with a mean square value around 5 nm. The mean size of the Si-ncs for the sample deposited at 900 °C varies between 3 and 5.5 nm as shown in the histogram (inset from Figure 3), while for the sample deposited at 1000 °C, Si-ncs between 5 and 15 nm were measured (see inset in Figure 4). The results are consistent with the average particle size calculated by the Scherrer formula (Table 1).

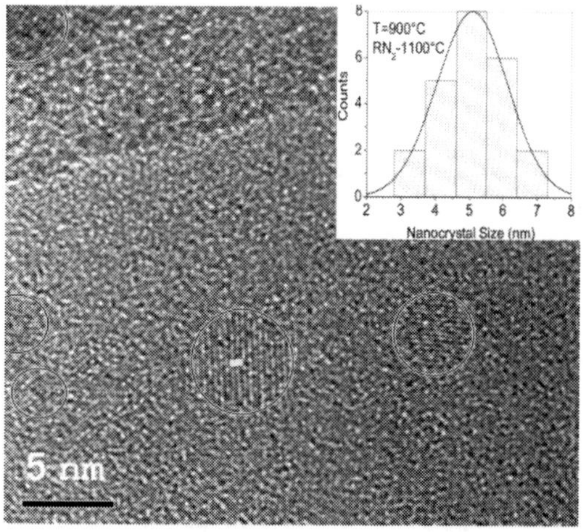

Figure 3. HRTEM image and histogram for the sample deposited at 900 °C with subsequent annealing at 1100 °C.

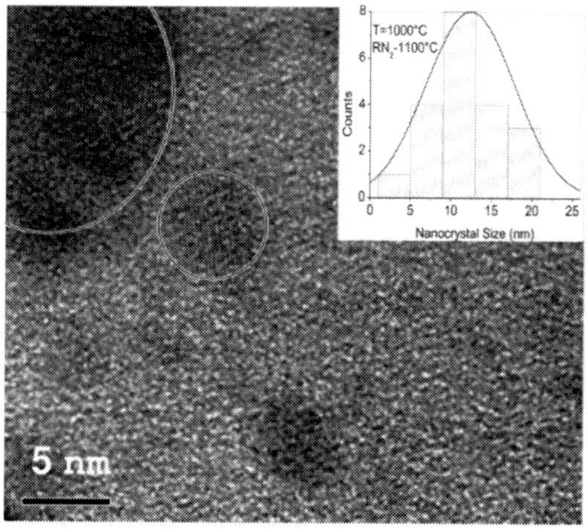

Figure 4. HRTEM image and histogram for the sample deposited at 1000 °C with subsequent annealing at 1100 °C.

Regarding the number of nanocrystals, HRTEM images of as-deposited samples were analyzed [11]. The average size particle found on the sample deposited at 900 °C was 3 nm and for sample at 1000 °C varied between 3 and 6 nm as shown the histograms. The HRTEM images showed greater number of Si-ncs compared to the as-deposited samples. We proposed that formation of Si-ncs is carried out entirely by diffusion mechanism of silicon in SiO_2 [12]. This diffusion is dependent on growth temperature. When the samples are annealed the diffusion coefficient of silicon in SiO_2 is higher and Si-ncs should therefore be larger. Furthermore, amorphous silicon remains present and crystallizes resulting in new Si-ncs. Thus, it can be confirmed that both the mean size and the number of Si-ncs is increased by effect of annealing process.

PL measurements were performed on SiO_x films at room temperature in the range of 400-1100 nm. In Figure 5, we compare the PL spectra of SiO_x films before and after annealing. It can be observed an increase in the PL intensity when the samples are annealed. The peak around 930 nm increases in intensity and undergoes a slight shift toward higher wavelengths. We propose that this peak is associated with radiative defects [11]. The increase in the number of Si-ncs with annealing is related to defects: an increase in the number of nanocrystals causes more radiative defects in this wavelength.

On the other hand after annealing, the band around 600-847 nm increases in intensity due to the large amount of nanocrystals formed, this band corresponds to nanocrystal sizes between 3 and 7 nm observed by HRTEM. In this case the model of quantum confinement might explain the radiative emission in this range of energy. We propose that both effects must be responsible for the PL phenomenon in our samples. However, a thorough study is needed to support these assumptions.

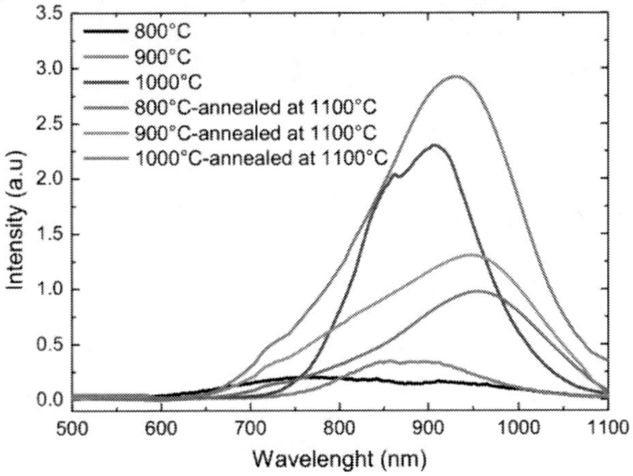

Figure 5. PL spectra of SiO_x films before and after annealing.

4. Conclusion

The effect of annealing on structural properties of SiO_x films obtained by HWCVD technique was studied. The different analysis showed an increase in the crystallization of silicon clusters when the samples were annealed. In both by HRTEM and XRD the size and the number of Si-ncs increases compared with the as-deposited samples. PL measurements in SiO_x films revealed an intense photoluminescence in the samples after annealing due to an increase in both the number of nanocrystals and increased of radiative defects.

References

[1] Lu Z H, Lockwood D J, and Baribeau J M 1995 *Nature* **378** 258.
[2] Shimizu I T, Nakao S and Saitoh K 1994 *Appl. Phys. Lett* **65** 1814.
[3] Mirabella S, Di Martino G, Crupi I, et al. 2010 *J. Appl. Phys.* **108** 093507.
[4] Nozik J 2008 *Chem. Phys. Lett.* **457** 3.
[5] Yuan Z, Pucker G, Marconi A, et al. 2011 Sol. Energy Mater. Sol. Cells. **95** 1224.
[6] Anbarasan P M, Senthilkumar P, Manimegalai S et al. 2010 *Silicon* **2** 7.
[7] Conibeer G, Green M, Corkish R, et al. 2006 *Thin Solid Films* **511–512** 654.
[8] Mota E, Meléndez M A, et al. 2010 *J. Appl. Phys* **108** 094323.
[9] Zhang X W 2005 *Phys. Status Solidi A* **202** 1773.
[10] Matsumoto Y, Godavarthi S, Ortega M, et.al. 2011 *Thin Solid Films* **519** 4498.
[11] Coyopol A, et al. 2012 *Journal of Nanomaterials*, ID 368268, 7 pag. doi:10.1155/2012/368268.
[12] Nesbit L A. 1985 *Appl. Phys. Lett.* **46** 38.

Low temperature ZnO films deposited by SSCVD

G Escalante[1], H Juárez, M Pacio, G García, T Díaz, E Rosendo and A Coyopol
Centro de Investigación en Dispositivos Semiconductores, Universidad Autónoma de Puebla, 14 Sur and Av. San Claudio, San Manuel 72570, Puebla, México.

[1]E-mail: gertrova@ece.buap.mx

Abstract: ZnO films were deposited by single source chemical vapor deposition (SSCVD) at low temperatures on glass substrates using zinc acetate as a source of Zn and oxygen gas as the oxidizing agent. The zinc acetate was sublimated on a resistance heated to 210°C. The substrate temperature was varied from 200 to 300°C every 50°C with a deposition time of 5 min. X-ray diffraction results show that all deposited films are polycrystalline with an average grain size below 8 nm in all directions. The ZnO films showed a transmittance higher than 75% in the visible region.

1. Introduction

Zinc oxide (ZnO) is an n-type semiconductor with a wide band gap (3.3 eV) and a hexagonal wurtzite structure [1]. The ZnO films are widely used for a variety of applications such as transparent conductive oxide (TCO) [2], surface acoustic wave device (SAW) [3] and gas sensor [4] due to their structural, optical and electrical properties. So far, a variety of techniques such as sputtering [5], sol-gel [6], spray pyrolysis [7] and atmospheric pressure chemical vapor deposition (APCVD) [8, 9] have been used to deposit ZnO films. Over the past few years, there has been a growing interest in single-source chemical vapor deposition (SSCVD) because of its low-cost deposition system and high-quality films at moderate temperatures. This technique essentially consists of the precursor vaporization followed by the precursor decomposition/deposition [10-13].

In this paper, we report the low temperature deposition of ZnO films obtained with an SSCVD system at atmospheric pressure. Also, structural and optical characterizations of the deposited ZnO films are presented.

2. Experimental Procedure

Figure 1 shows a schematic diagram of the SSCVD system used for the deposition of ZnO films. The system consists of a vertical reactor made of quartz; it was implemented in the laboratory at Centro de Investigaciones en Dispositivos Semiconductores [14]. The ZnO films were deposited on glass substrates using dihydrate Zinc acetate ($Zn(CH_3COO)_2 \cdot 2H_2O$), as a single precursor which was sublimated on a NiCr resistance heated to 210°C using an AC power source (10 Vp, 60 Hz). A type-K thermocouple and a Eurotherm 2416 temperature controller were used to control the resistance temperature. The resistance was placed at 50 mm below the substrate. Additionally, an Oxygen gas (O_2) flow was used as oxidant agent and the flow rate was 50 sccm. On the other hand, the substrate was placed under the upper heater. The substrate temperature (T_s) during deposition was varied between 200°C and 300°C every 50°C with a deposition time of 5 min.

IOP Conf. Series: Materials Science and Engineering **45** (2013) 012012 doi:10.1088/1757-899X/45/1/012012

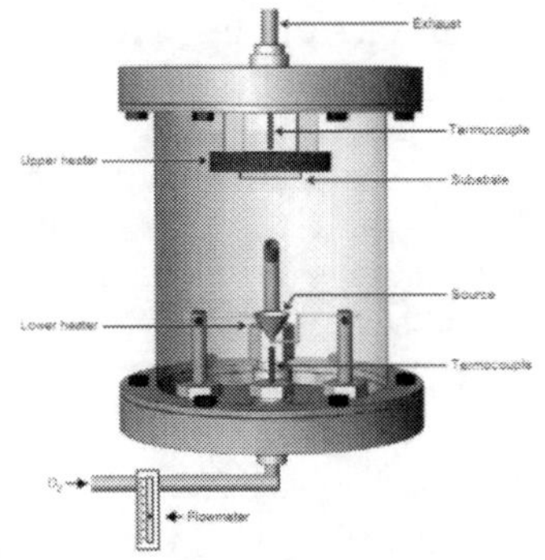

Figure 1. Schematic diagram of the SSCVD system.

The structural properties of the deposited ZnO films were studied using X-ray diffraction (XRD) in a Bruker D8 Discover diffractometer with a X-ray source of CuKα radiation (1.5406 Å) at 40kV and 40mA. The optical characterization was carried out using a UV-vis Thermo Scientific Evolution 600 spectrophotometer in the wavelength range from 200 to 800 nm. The thickness of the films was determined using an electromechanical stylus instrument.

3. Results and Discussion

Figure 2 shows the X-ray diffraction patterns of the deposited films. At a low deposition temperature of 200°C, the film is polycrystalline and without preferential orientation. The diffraction peaks were observed in 31.68°, 34.34° and 36.16° corresponding to the (100), (002) and (101) planes, respectively [15]. However, there is a contribution when increasing the temperature to 250°C; the deposited film has a preferred orientation in the (002) plane. Many researchers assume that the contribution is due to the low surface free energy in comparison to the (110) and (100) planes [16]. Nevertheless, the deposited ZnO film at 300°C is polycrystalline without preferential orientation. At this temperature, the adhesion of particles increases from the gaseous phase reducing the oriented growth.

The grain size (D) was calculated using the Scherrer formula [17]

$$D = \frac{0.9\lambda}{B\cos\theta} \tag{1}$$

where λ is the wavelength (Cu Kα = 1.5406 Å), θ is the Bragg's diffraction angle and *B* is the full width at half maximum (FWHM) of the diffraction peak. The inset of figure 2 shows that the values for the grain size obtained from the ZnO films are smaller than 8 nm. In addition, this shows that the grain size increases with the rise in the deposition temperature.

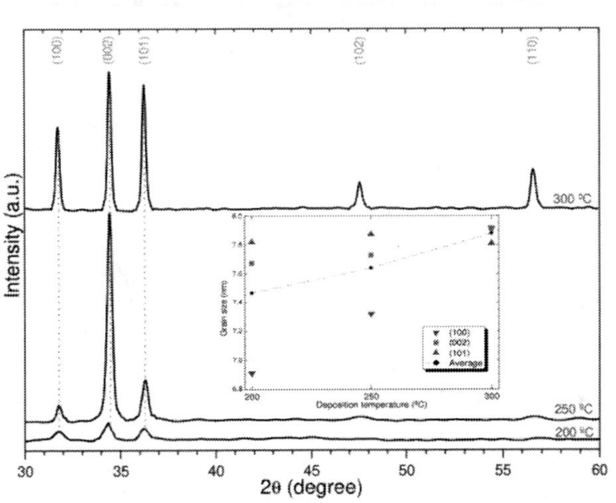

Figure 2. X-ray diffraction pattern of ZnO films deposited at different temperatures.

Figure 3 shows the transmittance spectra of deposited ZnO films in the temperature range from 200 to 300°C. The transmittance spectrum shows that all films have values higher than 75% in the visible region. The transmittance spectrum for the film deposited at 200°C shows a shoulder near to the UV region, this behavior can be related to the thin thickness and porosity of the film. Although the thickness increases in the case of the deposited films at 250 and 300°C, the percent transmittance in the visible region is maintained and the transmittance in the UV region disappears. This behavior is due to the fact that the film is densified when the temperature is increased [18].

The optical band gap of the deposited films was calculated using the relation

$$\alpha h\nu = A(h\nu - E_g)^{n/2} \tag{2}$$

where α is the absorption coefficient, $h\nu$ is the photon energy, A is a constant, E_g the optical band-gap energy and n = 1 (for semiconductor of direct transition) [19]. In the inset of figure 3, the relationship between $(\alpha h\nu)^2$ and $h\nu$ is plotted. The E_g value can be obtained by extrapolating the linear portion to the energy axis. The measured optical band gap values were in the range of 3.21 to 3.28 eV which is very close to the band gap of intrinsic ZnO [1].

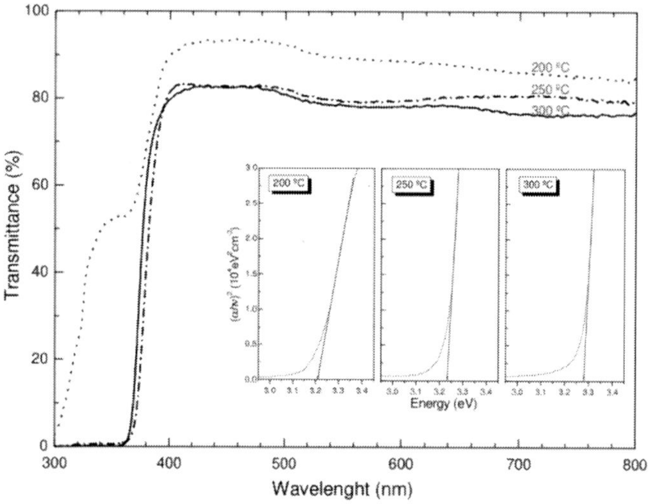

Figure 3. Transmittance spectra and plots of $(\alpha h\nu)^2$ as a function of the energy (E) for ZnO films deposited at 200, 250 and 300°C with thicknesses of 270, 380 and 690 nm, respectively.

4. Conclusions

ZnO films were deposited on glass substrates using the SSCVD technique at low temperatures. The films were deposited in the range from 200 to 300°C without treatment annealing. The ZnO film deposited at 250°C shows a greater crystalline quality associated with a high preferential orientation in the (002) plane. When the deposition temperature rises, the grain size increases. The optical transmittance in the visible region of the ZnO films shows that all films have a transmission (over 75%). Finally, it is shown that the percent of transmittance in the visible region of the deposited ZnO films depends on the films thickness and their densification.

Acknowledgment

The authors wish to acknowledge M.C. René Pérez Cuapio for his assistance in the realization of the X-ray diffraction.

IOP Conf. Series: Materials Science and Engineering **45** (2013) 012012 doi:10.1088/1757-899X/45/1/012012

References

[1] Özgür Ü *et al* 2005 *J. Appl. Phys.* 98 041301
[2] Jeong S H, Park B N, Yoo D G, Boo J H and Jung D 2007 *J. Korean Phys. Soc.* **50** 3 622-25
[3] Chen X, Liu D, Chen J and Wang G 2009 *Smart Mater. Struct.* **18** 115021
[4] Malik A, Sêco A, Nunes R, Vieira M, Fortunato E and Martins R 1997 *Mat. Res. Soc. Symp. Proc.* **471** 47
[5] Nunes P, Costa D, Fortunato E and Martins R 2002 *Vacuum* **64** 293-97
[6] Ohyama M, Kozuka H and Yoko T 1997 *Thin Solid Films* **306** 78-85
[7] Goyal D J, Agashe C, Takwale M G, Marathe B R and Bhide V G 1992 *J. Mater. Sci.* **27** 4705
[8] Haga K, Katahira F and Watanabe H 1999 *Thin Solid Films* **343-344** 145-47
[9] Pacio M, Juárez H, Escalante G, García G, Díaz T and Rosendo E 2010 *Mater. Sci. Eng. B* **174** 38-41
[10] Dai L P, Deng H, Zang J D, Mao F Y, Chen J J and Wei M 2008 *J. Mater. Sci.* **43** 312-15
[11] Deng H, Gong B, Petrella A J, Russell J J and Lamb R N 2003 *Sci. China Ser. E* **46** 4 355-60
[12] Deng H, Russell J J, Lamb R N, Jiang B, Li Y and Zhou X Y 2004 *Thin Solid Films* **458** 43-46
[13] Petrella A J, Deng H, Roberts N K and Lamb R N 2002 *Chem. Mater.* **14** 10 4339-42.
[14] Escalante G 2010 Design and fabrication of SSCVD system CIDS-UAP México
[15] International Centre of Diffraction Data, PDF: 00-036-1451
[16] Nguyen H T, Hartmann A J and Lamb R N 1999 *J. Phys. Chem. B* **103** 4264
[17] Klug H P, Alexander L E 1974 X-ray Diffraction Procedures for Polycrystalline and Amorphous Materials, (John Wiley & Sons. New York) p.p 687.
[18] Aslan M H, Oral A Y, Mensur E, Gül A and Basaran E 2004 *Sol. Energy Mater. Sol. Cells.* **82** 543-52
[19] Davis E A, Mott N F 1970 *Philos. Mag.* **22** 903-16.

Understanding the thermal annealing process on metallic thin films

G M Alonzo-Medina[1], A González-González[2], J L Sacedón[2] and A I Oliva[1]

[1]Centro de Investigación y de Estudios Avanzados del IPN Unidad Mérida, Departamento de Física Aplicada, AP-73-Cordemex, 97310 Mérida, Yucatán, México
[2]Instituto de Ciencia de Materiales de Madrid, Consejo Superior de Investigaciones Científicas, 28049 Madrid, Spain

Email: galonzo@mda.cinvestav.mx

Abstract. Thermal annealing is an usual process used for intrinsic stress liberation, structural improving, and surface roughness control in materials. In a qualitative way, annealing modifies the surface morphology of materials with temperature and time. In this work, a methodology to explain the surface modification of thin films when they are submitted to an annealing process is discussed. Two thermally evaporated Au films with 200 nm-thicknesses were post-annealed in a vacuum chamber at 100 °C with an argon atmosphere, and annealing times from 0 to 1000 minutes. Each Au film grown at different rate deposition was cyclical annealed and imaged under different annealing times. Data obtained from high quality AFM images after different annealing times of Au samples were used to calculate surface parameters such as roughness, grain size, and slope at the border, also the respective exponents as a function of the annealing time. The experimental results allow understanding the temporal evolution of the annealing process, as a rearrangement of the surface protrusions.

1. Introduction

Different analytical models based on intrinsic stresses and relaxation processes have been proposed to explain the surface evolution during the growth of metallic thin films [1]. Nevertheless, the occurring phenomena during thermal annealing have not been completely understood. Early studies explain that, during thermal annealing of Au thin films, plastic and elastic effects produce the partial collapse of the boundary network of surface protrusions (SPs), promotes relaxation processes driven by surface diffusion, diffusion into grain boundaries, and bulk diffusion [2].

In a previous work on Au(111) films [3] studied by atomic force microscopy and x-ray diffraction, authors explain the surface evolution by means of a continuum model that combines different mechanisms of stress (grain zipping and shear stress) and their correspondent mechanisms of relaxation (gap filling and grain rotation) with the interactions occurring between surface grains or surface protrusions. In those studies, different Au films samples were used for each annealing process.

In this work we proposed a quantification of the annealing phenomena by means of a statistical analysis of the temporal evolution of the surface parameters such as the lateral size, the slope at the border, and the surface roughness of the surface protrusions estimated from high-resolution AFM images. The same Au film sample was thermally annealed and imaged after each annealing process.

2. Experimental process and image analysis

In this section, the process to obtain the annealed Au thin films and the image analysis are detailed. This section is divided in four sub-sections. First, the films preparation is explained. Secondly, the methodology to calculate the slope distributions from AFM images is described. Thirdly, the methodology to obtain the grain size is discussed; and finally, the relations to estimate the rms-roughness values from surfaces are highlighted.

2.1. Thin film preparation

Two Au thin films (A and B) were deposited by thermal evaporation on Si substrates heated at ≈ 85 °C. During evaporation the substrate heats by radiation reaching an oscillating temperature between 80 and 95 °C. For this reason, the initial condition (as-grown) of samples A and B slightly differs between them. Otherwise, A and B samples were deposited separately under similar conditions. The pressure chamber during growth was 10^{-7} mbar. Au (99.999% purity) on SiO_x/Si (111) substrates was deposited. The p-type Si substrates have an electrical resistivity of 0.025 Ω-cm. The films were deposited at a growth rate of ≈ 1 nm/s, and a thickness of 200 nm. Au films were annealed in a vacuum oven with Argon atmosphere. The annealing temperature for all samples was 100 °C but annealing time ranged from 0 (as-deposited) to 1000 minutes. In this work, the annealing process was applied on the same sample for different annealing times. After each annealing time, a high resolution and a high quality atomic force microscopy image was taken. All AFM images were obtained into to the same zone of the film. This process is repeated after each annealing time (accumulated annealing time), which produces a cyclic thermal annealing on the Au thin film. For obtaining the surface images, an atomic force microscope (AFM) was used in tapping mode. The AFM images were obtained with 512 x 512 pixels^2 of resolution by using high aspect ratio tips with radius of 2 nm. Each AFM image was 2 x 2 μm^2 size.

2.2. Slope distributions

To obtain the slope distributions from the AFM images, a mathematical procedure consisting in a Lagrangian interpolation on the AFM image data matrix of the local heights and their conversion in a continuous function $h(\vec{r})$ (with $\vec{r}$ as vector position) was done. The local slopes $m(\vec{r})$ can be obtained by means of the equation:

$$m(\vec{r}) = tan\left(cos^{-1}\left(\frac{\vec{n}\cdot\hat{k}}{|\vec{n}|}\right)\right) \quad (1)$$

Where $\vec{n}$ is the normal vector to surface $\left(\vec{n} \equiv (n_x, n_y, n_z) \propto \left[\left(-\nabla_x h(\vec{r}), -\nabla_y h(\vec{r}), 1\right)\right]\right)$ and $\hat{k} = (0,0,1)$ the upward unit vector [4]. With these calculations the matrix of slopes and their slope distributions *N(m)* were obtained, being the maximum peak, the slope mean value.

2.3. Grain size of surface protrusions (SP).

In order to calculate the mean grain size λ (or lateral size) of the film surface, contour maps were obtained from the AFM images. Contour maps were obtained by means of the identification of the local minimum and saddle points of the $h(\vec{r})$ values. These last points are located into the periphery that delimits and surrounds the grains (grain boundaries). The grain size distributions were obtained by measuring the distances between intersections of the contour points in the x-y plane and assigning the maximum distance value to the SP lateral size. The maximum value in these distributions represents the mean grain size.

2.4. Rms-roughness values

The rms-roughness value w is the standard deviation or the fluctuations of the local heights of the AFM image. The expression to calculate w form AFM image data is given by:

$$w = \frac{1}{L}\sqrt{\sum_{LxL}\left(h(\vec{r}) - \langle h(\vec{r})\rangle\right)^2} \quad (2)$$

Where $\langle h(\vec{r})\rangle$ is the mean height, and L is is the window size of AFM image.

3. Results and Discussions

Figure 1 shows AFM images of the surface of the sample B annealed at different times. The images correspond to as-grown, 2, 20, 50, 200 and 1000 minutes of thermal annealing. Insets in figure 1 are the corresponding contour maps as obtained by the described process. In a similar way than the scaling laws proposed for the growing processes; as far as we know, we proposed for the first time, similar scaling laws of some involved parameters with time t, but for the annealing process. Figure 2 shows the behavior expressed as a power law, $w \propto t^{\gamma}$, of the surface roughness as a function of the annealing time for two Au annealed samples. Each point in Figure 2 represents data obtained from the corresponding AFM image. Initial rms- roughness value of the samples depends on the experimental growth conditions (such as the substrate temperature during deposition). The two analyzed samples show a similar behavior with roughness-exponent values γ of samples A and B fitted as $\gamma_A = 0.014 \pm 0.013$, and $\gamma_B = 0.015 \pm 0.007$, respectively. The roughness values increase with the annealing time, meaning a vertical growth of the surfaces protrusions during the annealing process.

From figure 3, the grain size of the surfaces protrusions is plotted as a function of the annealing time. Contrarily to figure 2, a decrease of the grain size with the annealing time can be observed. By fitting this behavior as a power law, $\lambda \propto t^{\delta}$, the grain size-exponent δ values for samples A and B are estimated as $\delta_A = -0.019 \pm 0.012$, and $\delta_B = -0.020 \pm 0.007$, respectively. In this case, we can affirm that while the surface roughness (vertical size) increases, meanwhile the grain size (lateral size) decreases; that is, the surfaces protrusions grow vertically while their base (grain size) diminishes. Note the similar value of the exponents for both samples. The high dispersion reported for the exponents is understood if we take into account that each AFM image was not obtained on the same site of the sample after each annealing process. Similar dispersion values were reported for the slope in ref. [5] during Au growth.

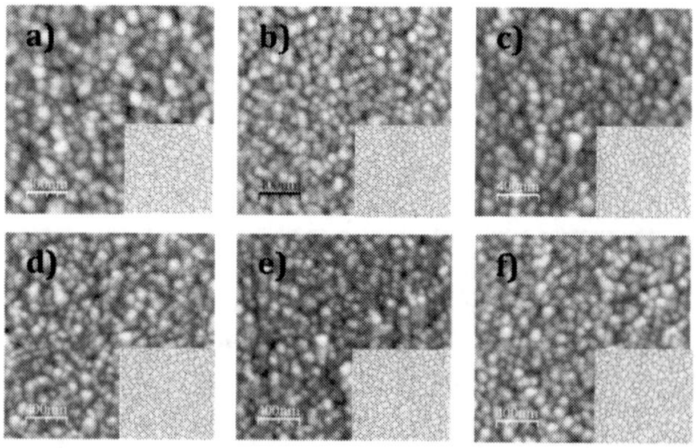

Figure 1. High-resolution AFM images of Au films (sample B) after different annealing process: a) as-grown, b) 2 min, c) 20 min, d) 50 min, e) 200 min, and f) 1000 min. Each inset is the corresponding contour map.

Figure 4 shows the evolution of the mean slope with the annealing time on the borders of surfaces protrusions. By fitting this behavior with annealing time as a power law, $m \propto t^{\varepsilon}$, the slope-exponents ε for samples A and B are estimated as $\varepsilon_A = 0.033 \pm 0.030$, and $\varepsilon_B = 0.027 \pm 0.010$, respectively. In good agreement with the roughness plot, the mean slope value increases with the annealing time as a result of the vertical growth of surface protrusions. In a complimentary way, Figures 5 and 6 show comparative plots between roughness and mean slope, as well as grain size and mean slope, respectively, as a function of the annealing time for sample A. From figure 5, it can be observed that roughness and mean slope parameters follow a similar behavior, affirming that these parameters have a similar behavior with annealing time. However, in figure 6, a contrarily behavior between the grain size and the mean slope with annealing time are observed. Similar behavior, but with different values were obtained for sample B (figures 7 and 8).

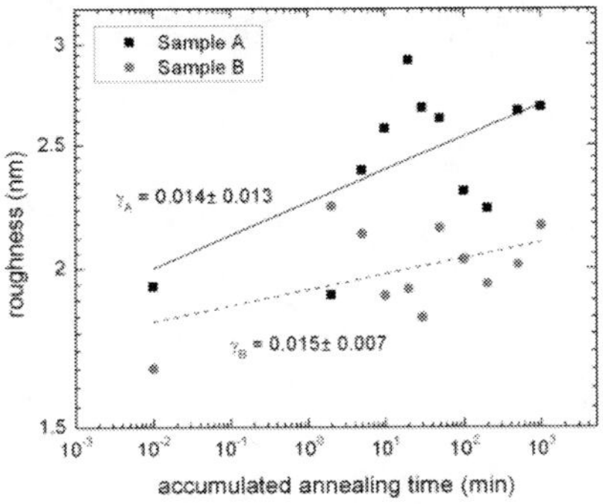

Figure 2. Log-log plot of roughness vs. accumulated annealing time. The annealing time ranged from 0 to 1000 min.

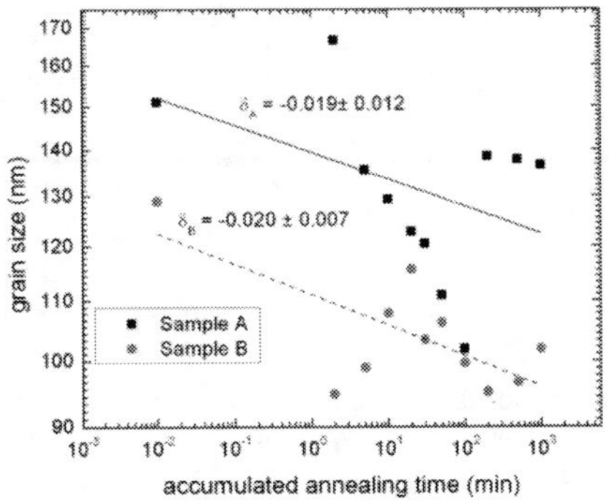

Figure 3. Behavior of grain size of surfaces protrusions with accumulative annealing time.

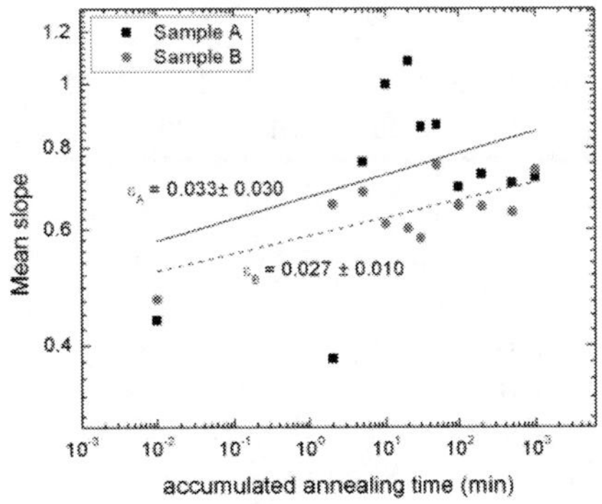

Figure 4. Mean slope at the borders of surfaces protrusions vs. annealing time.

Figures 7 and 8 show the plot of roughness and slope, and grain size and slope with the accumulated annealing time as obtained for sample B. A similar qualitative behavior of these parameters between sample A and sample B is observed. However, different absolute values of roughness, grain size and slope are reported for samples A and B, given the different growth conditions between the two gold samples.

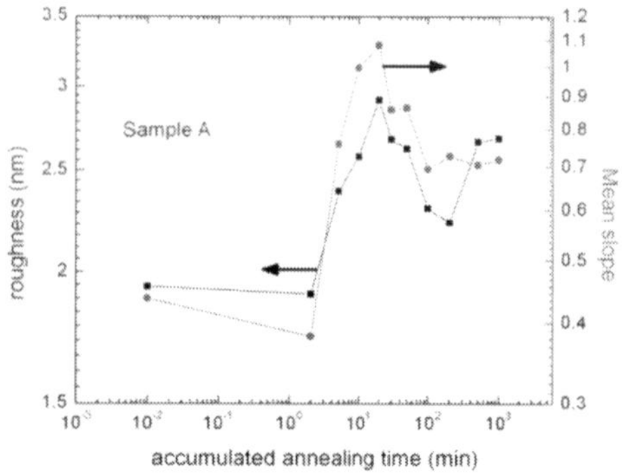

Figure 5. Rms-roughness and mean slope vs. annealing time as obtained for sample A. Parameters follows similar behavior.

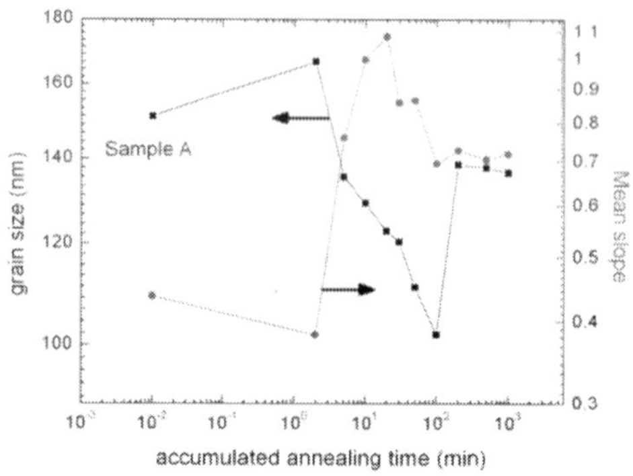

Figure 6. Grain size and mean slope vs. annealing time for sample A. Parameters show contrarily behavior.

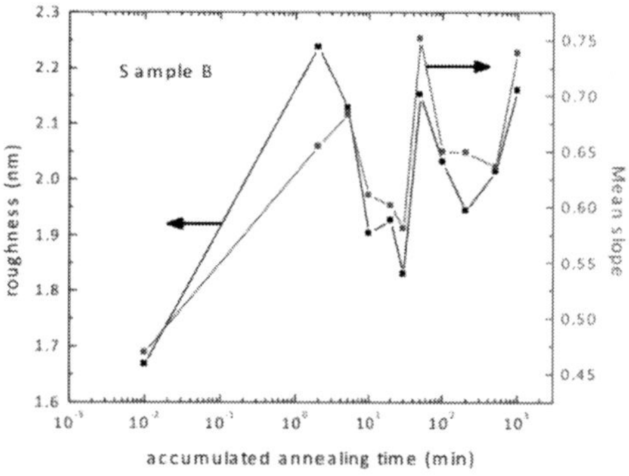

Figure 7. Rms-roughness and mean slope vs. annealing time as obtained for sample B. A similar behavior than sample A can be observed.

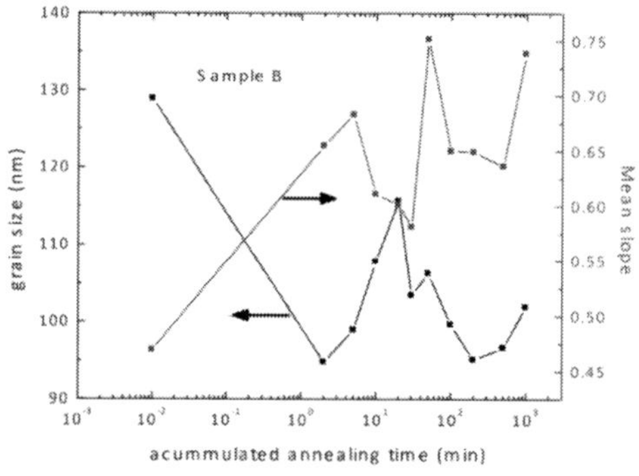

Figure 8. Grain size and mean slope vs. annealing time for sample B. A similar behavior than sample A can be observed.

In conclusion, our reported results are in good agreement with the objective of the annealing process: relaxation and intrinsic-stress liberation through surface protrusions reorganization with temperature. Furthermore, surface roughness w, grain size λ, and the slope m at the border grain parameters were quantified as a function of the annealing time from the analysis of the high quality AFM images taken on surface film after each annealing process. Scaling law exponents for each parameter were reported. Our results are in good agreement with reported works but for different materials. For example, from AFM images of FePt epitaxial thin films [6], authors report that the grain size reduces, meanwhile the magnetic properties improves when the films are annealed at 550 °C. In other work [7], the grain size decreases when $L1_0$ phase of FePt thin films deposited on oxidized Si (100) substrates are thermally annealed at 700 °C, at different heating rates (between 20 and 100 °C/s). In a recent work [8], a model to explain the surface morphology evolution by means of surface parameters as studied in this work, but using different films for each annealing process, was proposed.

References

[1] Freund L B and Suresh S 2003 *Thin Film Materials: Stress, Defect Formation and Surface Evolution* (England: Cambridge University Press, Cambridge

[2] Porath D, Goldstein Y, Grayevsky A, and Millo O 1994 *Surf. Sci.* **321** 81

[3] González-González A, Alonzo-Medina G M, Oliva A I, Polop C, Sacedón J L, and Vasco E 2011 *Phys. Rev. B* **84** 155450

[4] González-González A, Sacedón J L, Polop C, Rodríguez-Cañas E, Aznárez J A, and Vasco E 2009 *J. Vac. Sci. Technol A* **27** 1012

[5] Rodríguez-Cañas E, Aznárez J A, Oliva A I, and Sacedón J L 2006 *Surface Science* **600** 3110.

[6] Albertini F, Nasi L, Casoli F, Fabbrici S, Luches P, Rota A, and Valeri S 2007 *Journal of Magnetism and Magnetic Materials* **316** e158-e161

[7] Chen S C, Chen S p, and Kuo P C 2009 *Thin Solid films* **517** 5176-5180

[8] Alonzo-Medina G M, González-González A, Oliva A I, Sacedón J L, and Vasco E 2012 *J. Phys. D: Appl. Phys.* **45** 435301

IOP Conf. Series: Materials Science and Engineering **45** (2013) 012014 doi:10.1088/1757-899X/45/1/012014

Effect of the pos-deposition annealing with $CdCl_2$ on the optical, structural and morphological properties of CdTe-films grown by CSS

C Ricárdez-Jiménez[1], G Pérez-Hernández[1], J Pantoja-Enríquez[2], B Escobar-Morales[3], Q Angulo-Córdova[1], L L Díaz-Flores[1], M E Hernández-Torres[4], J M Gracia-Jiménez[4], N R Silva-González[4], X Mathew[5]

[1]*Universidad Juárez Autónoma de Tabasco, Avenida Universidad S/N, Zona de la Cultura, Col. Magisterial, Villahermosa, Centro, Tabasco 86040, México.*
[2]*Centro de Investigación y Desarrollo Tecnológico en Energías Renovables, UNICACH, Libramiento Norte No 1150, Tuxtla Gutiérrez, Chiapas 29039, México.*
.[3]*Instituto Tecnológico de Cancún, Av. Kábah Km. 3, Cancún, Quintana Roo 77500, México.*
[4]*Instituto de Física, Benemérita Universidad Autónoma de Puebla, Puebla, Puebla J-48 72570, México.*
[5]*Centro de Investigación en Energía Universidad Nacional Autónoma de México, Temixco, Morelos 62580, México.*

E.mail: rcristino@hotmail.com

Abstract. A set of CdTe films grown by the CSS method were annealed in a$CdCl_2$ atmosphere of during 5 min at 345, 365, 385, 405, 425, 435, 445 and 455 °C. The films were characterized before and after annealing by the SEM, EDS, X-ray, optical absorption and photoluminescence (PL) techniques. As the annealing temperature is increased a tendency to improve the Cd/Te ratio is observed. After the thermal treatment at 430 °C the films reach the stoichiometry. A deviation from stoichiometry appears again for annealing temperatures higher than 430 °C. The PL spectra of the films at 6.5 K exhibit two emission bands, one corresponds to the defects energy levels band (1.4-1.5 eV) and the other is located in the interval from 1.564 to 1.584 eV (pA). For annealing temperatures higher than 430 °C an increment in the intensity and energy position of pA is observed, while for annealing temperatures less than 430 °C the behavior is irregular. In the emission mechanisms of pA the expected impurities and/or native defects in the material such as the V_{Te}, $V_{Cd}Cl_{Te}$, Cd_{Te} and Cl_{Te} are involved.

1. Introduction

CdTe is a material that has attracted growing interest due to their applications in solar cells, because it has a theoretical efficiency of 29 %, a direct energy band gap of 1.45 eV, and an optical absorption coefficient greater than 10^4 cm^{-1} [1, 2]. CdTe/CdS solar cells with an efficiency of 16.5 % have been produced with polycrystalline CdTe thin films, but the efficiency in big modules is considerably smaller [3-5]. It is known empirically that, a pos-deposition thermal treatment of CdTe in an atmosphere of $CdCl_2$ (or HCl) improves the efficiency of the device, however the exact mechanism of the process is not yet completely known.The thermal treatment in an atmosphere of $CdCl_2$ is associated with a recrystallization, interdiffusion at the CdTe/CdS interface, and with a decrement of the series resistance [6-9]. Complete devices with the same pos-deposition annealingin a $CdCl_2$ atmosphere have reached efficiencies around 11.6% [10]. The simplest way to study the pos-deposition effects of $CdCl_2$ or Cu atmosphere on CdTe is when this material is deposited in the form of thin film on glass substrate. In this work, we study the effects of the $CdCl_2$ atmosphere on CdTe thin films by the techniques: scanning electron microscopy (SEM), energy dispersive spectroscopy (EDS), X-ray diffraction (XRD), optical absorption, and photoluminescence (PL).

2. Experimental

The CdTe films were grown on glass substrates by the close spaced sublimation (CSS) method with the following deposition conditions: pressure (10 mbar), source temperature 600 °C, substrate temperature 550 °C, carrier gas proportion O_2:He (1:4), source and substrate separation of 2 mm, and deposition time of 3 min. For the pos-deposition thermal treatment a source of $CdCl_2$ (99.99%) was employed using a system similar to the one used in the CdTe growth process. The annealing of the CdTe films was carried out for 5 min at the temperatures: 345, 365, 385, 405, 425, 435, 445 and 455 °C, while the CdCl2 source temperatures were maintained at 350, 370, 390, 410, 430, 440, 450 and 460 °C, respectively.

The SEM and EDS characterization was performed with a JEOL (JSM-5400LV) system coupled with a Noran (Quest) spectrometer. The structure of the films was studied using a Rigaku X-ray diffractometer using CuK_α radiation (1.54056 Å). The absorption spectra were measured at room temperature using an IR interferometer BRUKER-EQUINOX 55. For the PL analysis, the samples were placed in a cryogenic refrigerator and the spectra were recorded at 6.5 K. The sample was excited with the 488 nm line of an Ar^+ laser. A SPEX 1000 M monochromator with ± 1 meV resolution was used for the spectral analysis. A GaAs photomultiplier was used for signal detection and the data were processed using a DATASCAN computer-controlled system.

3. Results and discusion

Figure 1 shows the surface morphology at X7500 magnification of the as-grown CdTe film, and of the annealed films at 350, 430, and 460 °C, which are representative of the samples set. As can be observed the CdTe films are polycrystalline, at the higher annealing temperatures the grain boundaries tend to disappear, a partial fusion of the grains is observed with a consequently increment on the grains size. This behavior is in accord with previous reports which point out that the recrystallizationis enhanced by the $CdCl_2$ source and temperature [10], i.e. as the annealing temperature increases the Cd content is increased in the films, promoting in this way the fusion and growth of the grains. This proposal is confirmed by the EDS analysis.

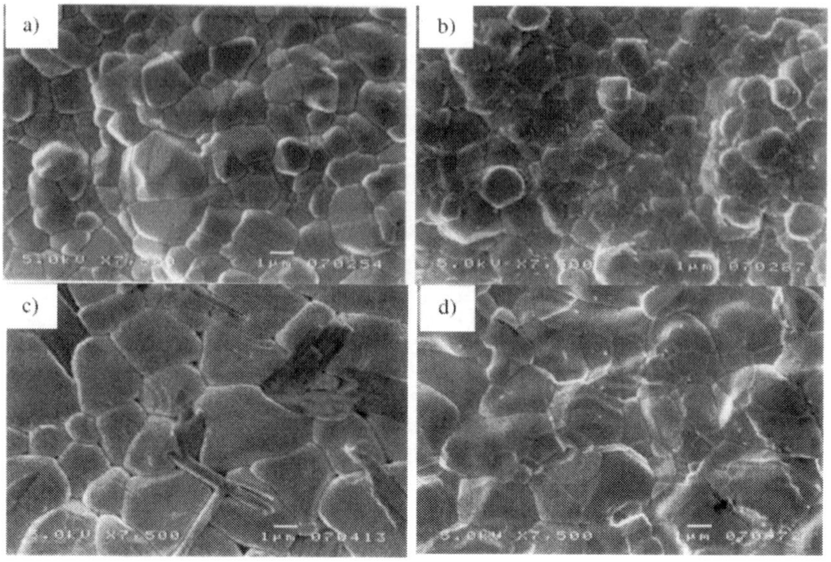

Figure 1. Surface morphology of the CdTe films: a) as-grown, b), c), and d) annealed in a $CdCl_2$ atmosphere at 350, 430, and 460 °C, respectively.

Figure 2 shows the average atomic percentage ratio Cd/Te of the CdTe films annealed in a $CdCl_2$ atmosphere at different temperatures. As it can be observed the atomic ratio presents a maximum, which coincides with the stoichiometric value of CdTe [11], and it corresponds to the CdTe film annealed at 430 °C. The rest of the films have a deviation from stoichiometry with Te excess. As the source temperature of $CdCl_2$ increases the Cd incorporation in the films also increases, compensating the Te excess and reaching the stoichiometry. This confirms the proposal of the SEM analysis. However, for annealing temperatures higher than 430 °C the atomic ratio Cd/Te of the films decreases again, and the films once again present a Te excess. This behavior is probably due to the conditions of pressure, carrier gas flux and annealing temperature, which promote the Cd evaporation from the films, even though the recrystalization process continues.

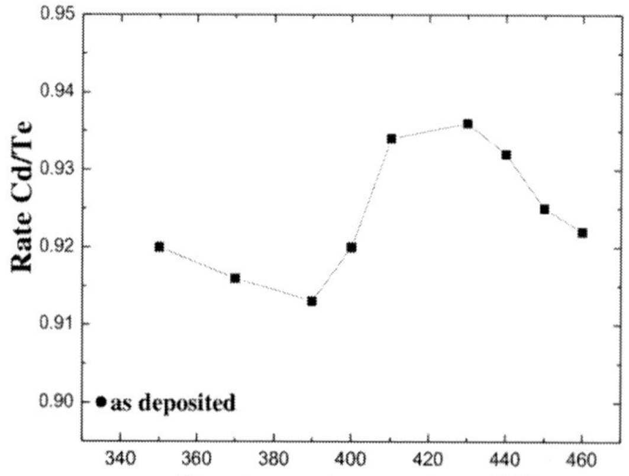

Figure 2. Evolution of the average atomic percentage ratio Cd/Te vs annealing temperature in an atmosphere of $CdCl_2$.

The X-ray diffraction patterns of the CdTe films annealed in a $CdCl_2$ atmosphere at different temperatures are shown in Fig.3. The spectra were obtained by scanning 2θ in the range 20–78°, with a grazing angle equal to 1.0°. The XRD pattern shows the characteristic peaks of the cubic zincblende structure and the preferential orientation of the grains for the (111) plane. The CdTe samples have a cubic zinc-blend structure with a strong preferred orientation along the (111) planes parallel to the substrate. For the samples annealed at T>450 °C the formation of oxides of cadmium and tellurium such as CdO, TeO_x and $CdTe_yO_x$ is evident as we can see from the spectra (Figure 3 with highlighted arrows), this is due to the system is kept in an atmosphere of He_2+O_2 [13]. In the case of samples annealed at T<430 °C the oxide peaks were not observed].

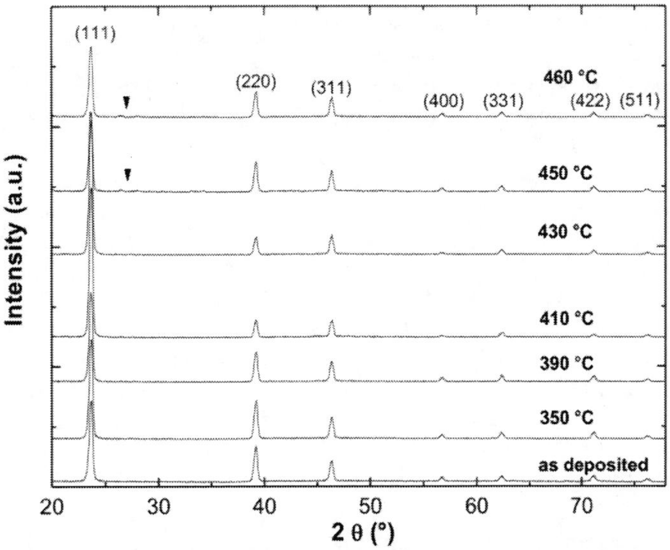

Figure 3. XRD pattern of the CdTe films annealed in a $CdCl_2$ atmosphere.

The lattice parameter, $\mathbf{a_o}$, of the samples were calculated from the peak positions in XRD patterns and using the method developed by Nelson and Taylor [13, 14]. The lattice parameter a_o of different CdTe films annealed in a CdCl2 atmosphere at different temperatures are shown in Fig. 4. For the as-deposited samples, the value of $\mathbf{a}_o$ ($\mathbf{a}_{as\text{-}deposited}$ = 6.479 Å) is less than the powder sample ($\mathbf{a}_{powder}$ = 6.481 Å), which suggests that the as-deposited film is submitted to a tensile stress in the plane parallel to the substrate surface. This stress is caused by the lattice mismatch and differences in thermal expansion coefficients between the CdTe film and substrate. It can be seen that the lattice parameter has a dependence on temperature, the lattice value decrease and reaches a minimum value at T=370 °C, then increases reaching a maximum value at temperature between 390 and 410 °C and then again decrease to reach a minimum value (6.471 Å) at T=470 °C. This behaviour suggests that the films throughout heat treatment are under tensile stress. In the figure, the first point corresponds to the as deposited film. The dotted straight horizontal line corresponds to the stress free value of the lattice constant.

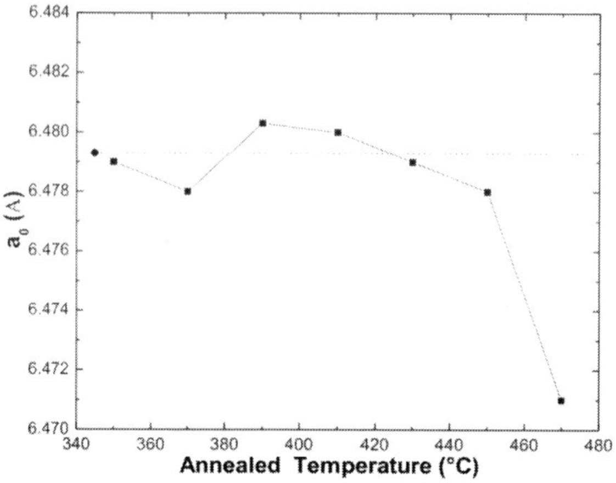

Figure 4. Lattice parameter $\mathbf{a_o}$ of the CdTe films annealed in a $CdCl_2$ atmosphere.

Figure 5 presents the energy of the optical band gap (Eg_{op}) of the CdTe films as a function of the annealing temperature. As it can be observed, all CdTe films have Eg_{op} values greater than 1.45 eV corresponding to the CdTe bulk. However, these values are similar to that reported for CdTe films grown by the same technique [15].This behavior could be explained by considering that the bottom of the conduction band levels arepartially filled with electrons coming from donor defects or impurities levels which have been thermally excited, and the acceptor defects or impurity levels get fill from electrons coming from the valence band top. So, the electron transitions happen at an Eg_{op} greater than the correspondent to CdTe bulk. On the other hand, as the annealing temperature increases an increment on the Eg_{op} values is observed, then a decrement and afterwards an increment is observed again. The minimum Eg_{op} value is obtained for the thermal treatment at 430 °C, and it corresponds to the stoichiometric film. As it was observed in the SEM and EDS results, the thermal treatments produce changes in the morphology and in the atomicratio, consequently the concentration of defects and impurities will be modify, therefore the carriers concentrations in the bands, the filling of the impurity levels and the Eg_{op}values will change as well [16]. By considering that, a stoichiometric CdTe film hasthe least possible defects and impurities concentration, it isreasonableto think that this situation corresponds to the CdTe compound with the smallest Eg_{op} value.

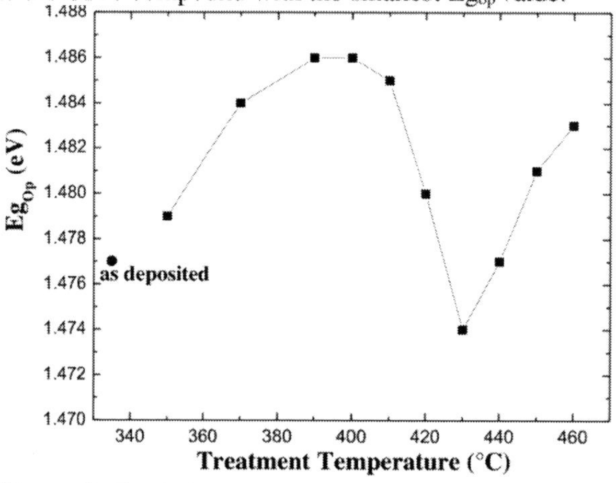

Figure 5. Evolution of the optical energy band gap (Eg_{Op}) vs annealing temperature in a $CdCl_2$ atmosphere.

The PL spectra measured at 6.5 K and constant excitation of the as-grown samples and of the thermal treated in aCdCl$_2$ atmosphere as well are shown in figure 6. Two bands, one in the region from 1.4 to 1.5 eV, which corresponds to the well known defects band of CdTe (DB) [17], and the other in the region close to the Eg$_{op}$ value (IB), are observed in the spectra.In general, the peaks position and intensity associated to the IB band present a minimum for the thermal treatment at 430 °C (see figures. 7a, 7b), while the DB band exhibits an irregular behavior. The DB band is conformed by the superposition of several transitions and phononic replies [18, 19], which involve defects and impurities, therefore the anomalous behavior can be explained by taking into account that, a variation of both defects and impurities concentration exists, and consequently the lines at low energy corresponding to recombination processes dominate for some temperatures, while for other temperatures the lines at high energy dominate. In this competing process the peaks will shift according to the domination degree. A similar situation occurs for the IB band, but in this case the spectra correspond to isolated lines, since shoulders can not be observed at first sight, i.e. only one type of impurity is involved. The identified emission mechanism and the type of impurities and complexes involved as well as listed in Table 1 indicate that the pos-deposition thermal treatments modify their concentration continuously.

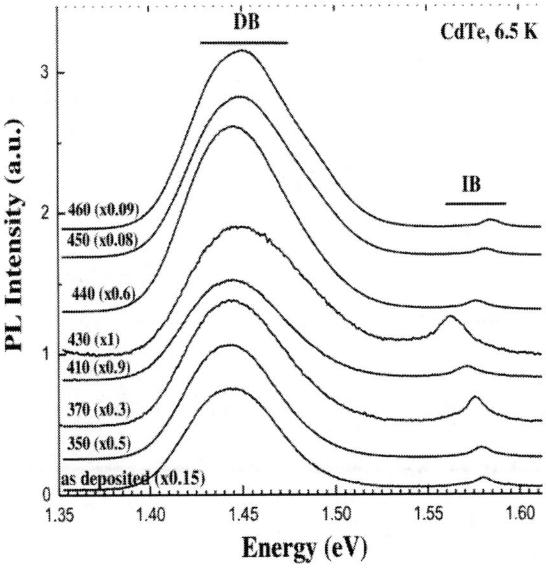

Figure 6. PL spectra of the CdTe films annealed in a CdCl$_2$ atmosphere at 6.5 K.

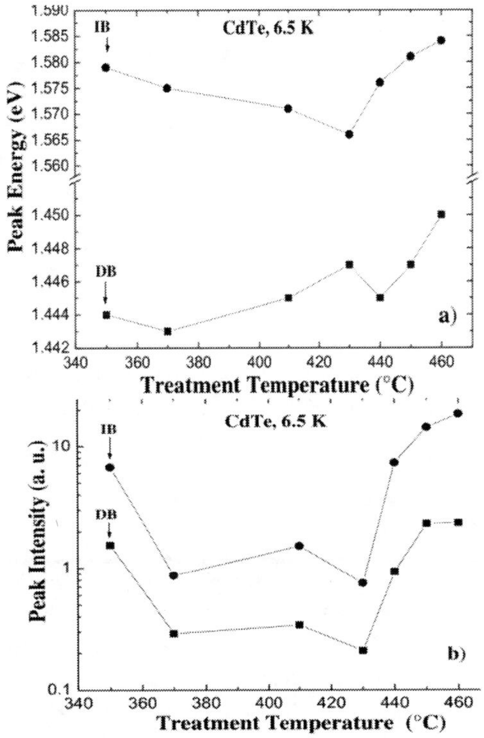

Figure 7. Evolution of the peaks position (a) and peaks intensity (b) of photoluminescence vs annealing temperature in a CdCl$_2$ atmosphere

Table 1. Emission mechanism and the type of impurities and complexes involved as well.

Sample	Peak Energy (eV)	Ionization Energy (eV)	Impurity/Defect	Transition
ST	1.580	0.021	V_{Te}^{+} (Donor)	I-B [2,15]
350	1.579	0.022	V_{Te}^{+} (Donor)	I-B [2,15]
370	1.576	0.025	Impureza	I-B or B-I [18]
410	1.571	0.029	V_{Cd}-Cl_{Te} (Aceptor)	B-I [18]
430	1.564	0.037	-----	-----
440	1.576	0.025	Impureza	I-B or B-I [18]
450	1.581	0.020	V_{Te}^{+} (Donor)	I-B [2,15]
460	1.584	0.017	Cl_{Te}^{+} (Donor)	I-B [20]

4. Conclusion

The results show that the pos-deposition annealing in an $CdCl_2$ atmosphere modify the morphologycal, structural, compositional and optical properties of the CdTe films. So, the annealing process allows to control the energy of the optical band gap, the Cd/Te atomic ratio, the emission mechanisms, and the defect and impurity concentrations. The CdTe film annealed at 430 °C exhibits properties close to that of monocrystalline CdTe with low defect and impurity concentrations.

Acknowledgments

Financial support from the Universidad Juárez Autónoma de Tabasco (UJAT), Fondo Mixto-Tabasco (FOMIX-Tabasco), Consejo Nacional de Ciencia y Tecnología (SEP-CONACYT) and Programa del Mejoramiento del Profesorado (PROMEP) is gratefully acknowledged through the project CB-2007/83960, TAB-2009-C18-121587, UJAT-PTC-144 and UJAT-2009-C05-60.

5. References

[1] Durose K, Edwards P R, Halliday D P 1999 J. Crystal Growth **197** 733.
[2] Zanio K 1978 Cadmium Telluride (Ser. Semiconductors and Semimetals Vol. **13**), Academic Press, New York, San Francisco, London.
[3] Britt J, Ferekides C 1993 Appl. Phys. Lett. **62** 2851.
[4] Moutinho H R, Hasoon F S, Abulfotuh F, Kazmersky L L 1995 J. Vac. Sci. Technol. A Vac. Surf. Films **13** 2877
[5] Wu X, Keane J C, Dhere R G, DeHart C, Albin D.S, Duda A, Gessert T A, Asher S, Levi D H, Sheldon P 2001 Proceedings of the 17th European PVSEC, Munich Germany, pp. 995-1000.
[6] Basol B, Ou S, Stafsudd O 1985 J. Appl. Phys. **58** 3809.
[7] Al-Allak H, Brinkman A, Richter H, Bonnet D 1996 J. Cryst. Growth **159** 159.
[8] Jahn U, Okamoto T, Yamada A, Konagai M 2001 J. Appl. Phys. **90** (2001) 2553.
[9] Paulson P, Dutta V, 2000 Thin Solid Films **370** 299.
[10] Al-Jassim M M, Yan Y, Moutinho H, Romero M J, Dhere R D, Jones K M 2001 Thin Solid Films 3**87** 246.
[11] Aven M, Prener J S 1967 Physics and Chemistry of II-Vi Compounds, North-Holland publishing company-Amsterdam.
[12] Hernández-Torres M E, Silva-González R, Casarrubias-Segura G, J M Gracia-Jiménez J M 2006 Sol. Energy Mater. Sol. Cells. **90** 2241.
[13] Moutinho H R, Hasoon F S, Abulfotuh F, Kazmersky L L 1995 J. Vac. Sci. Technol. A. **13** 2877.
[14] Taylor A and Sinclair H 1945 in Proceedings of Physical Society, Vol. **57** (The Physical Society, London) pp. 126.

IOP Conf. Series: Materials Science and Engineering **45** (2013) 012014 doi:10.1088/1757-899X/45/1/012014

[15] Nelson J B and Riley P 1945 Proc. Phys. Soc. **57** p.160
[16] Hernández-Torres M E 2006 "Estudio de selenuro de cadmio crecido por depósito en baño químico: Crecimiento y caracterización", Thesis Ph.D. BUAP p. 65.
[17] Feng Z S, Mascareñas A, Choyke W J, Farrow R F, Shirland F A and Takei W J 1985, Appl. Phys. Lett. **47** 24.
[18] Espinoza J E, Gracia J M, Navarro H, Zehe A 1983 J. Lumin. **28** 163.
[19] Myers T H, Schetzina J F, Edwards S T, Schreiner A F 1983 J. Appl. Phys. **54** 4232.
[20] Mathew X, Arizmendi J R, Campos J, Sebastián P J, Mathews N R, Ricardez-Jiménez C, Jiménez M G, Silva-González R, Hernández-Torres M E, Dhere R 2001 Sol. Energy Mater. Sol. Cells. **70** 379.

IOP Conf. Series: Materials Science and Engineering **45** (2013) 012015 doi:10.1088/1757-899X/45/1/012015

Influence of variation in the concentration of ammonium hydroxide on the size of ZnO crystal obtained by Microwave Chemical Bath Deposition

R. Galeazzi[1*], T. Díaz[1], G.García[1], B.L. Rivera[1], E. Rosendo[1], R. López[1], N. Morales[1] and C. M. González[2]

[1]CIDS-ICUAP, Benemérita Universidad Autónoma de Puebla, 14 Sur y Av. San Claudio s/n, C.U. Col. San Manuel, Edif. 136-C, Puebla, Pue., México. C. P. 72570. Tel. 01(222) 2295500 ext. 7876. [2]Facultad de Ciencias Químicas, Benemérita Universidad Autónoma de Puebla, 14 Sur y Av. San Claudio s/n, C.U. Col. San Manuel, Edif. 105-C, Puebla, Pue., México. C. P. 72570.

[*]ingquim25mail.com

Abstract. Films of good crystalline quality of ZnO were successfully prepared using the microwave chemical bath deposition method at a temperature of 80 °C. Concentration of the basic precursor was varied systematically in order to obtain different degrees of acidity in the precursor solutions. Increasing the pH causes an increase in yield. This increase is reflected on the thickness of the deposit. The results of atomic force microscopy (AFM) show an increase in particle size with increasing pH in agreement with the results obtained by profilometry.

Introduction

The metal oxides are currently the vehicles to achieve multifunctional heterostructures, nanostructures, and multilayer and thin films that are the basis of new optical devices whose electrical and magnetic properties interact one other. Furthermore, the metal oxides have extraordinary properties such as high electronic conductivity, optical transparency, superconductivity, thermoelectricity, etc. Therefore, the knowledge in the production of films obtained by altering its physical and chemical properties and by changing the processes of production is extremely important for the development of new applications for these materials.

The importance of ZnO is specifically based on its wide direct band gap that is 3.37 eV at room temperature and the large exciton binding energy of 60 meV. The significance of the ZnO films is such that various techniques have been employed to obtain it, such as RF magnetron sputtering at various temperatures [1], thermal oxidation of Zn ion beam sputtered films [2] Zn-vapor release in argon atmosphere and subsequent oxidation in air [3], vapor-deposition method (PVD) [4], chemical vapor deposition (CVD) [5], etc.

It has also been reported the synthesis of nanoparticles of ZnO in aqueous medium under microwave irradiation conditions, deposited on different substrates from zinc salts and various solvents [6]. However, no other report related to the variation in pH has been found. Chung-Hsin Lu *et al.* [7], performed an analysis related to obtain ZnO precipitates varying the pH of the solutions under hydrothermal conditions.

In this work we present the study on the variation on the concentration of ammonium hydroxide on the size and crystalline quality of ZnO obtained by microwave chemical bath deposition (MWCBD). The importance on the study and characterization on the properties of the films obtained by varying the pH of precursor solutions is that, the results obtained might help us to a better understanding of the physicochemical changes of ZnO obtained under these growth conditions.

1. Material and methods

As result of the chemical reactions in this deposition technique using microwave radiation as a way of heating, the material was obtained as a film and as precipitated powders. Because there are some reports on the behavioral grain size of ZnO precipitated powders [8], in this report we only mention the results obtained which are related to films.

Zinc nitrate hexahydrate salts ($Zn(NO_3)_2{\cdot}6H_2O$, J. T. Baker) were used as a source for zinc ions, urea (CH_4N_2O, Sigma Ultra) was also used as a complexing agent, ammonium hydroxide (NH_4OH, E. M. Science) as oxidizing medium and Corning glass slides were used as substrates.

The solutions of the precursor reagents were prepared at room temperature using deionized water of 18.2 MΩ of resistivity with the purpose of diminishing the residual impurity quantity. Five solutions were prepared with molar ratio of 2:1 (0.2 M/0.1 M) of $Zn(NO_3)_2$/Urea, and NH_4OH was added to achieve pH of 9.25, 9.88,10.39, 10.94 and 11.14. Three glass substrates were immersed into each solution, and then, they were placed in the microwave oven. The initial heating time was 4 minutes at maximum power reaching a temperature of 80 °C, after this, heating was maintained at this temperature for 40 minutes, using the minimum power up to terminate the reaction. Finally, the films were rinsed with strong agitation and dried with gaseous nitrogen.

Structural characterization of the layers was carried out using X-ray diffraction (XRD). The measurements were performed on a Bruker D8 Discover diffractometer, with a geometry of parallel beams and a gobel mirror monochromator, The diffractometer worked with Cu Kα radiation = 1.5406 Å, a tube conditions of 40 KV an 40 mA, the measurement interval was from 20° to 80°, with a step size 0.02°. The topographical characterization was performed in an Atomic Force Microscope AMBIOS. The data were processed with the program WSxM 5.0 [9]. The thickness of the films was measured in a profilometer Dektak Stylus 150 with a diamond tip of 12.5 μm in diameter at the base at a resolution of 10 nm.

2. Results and discussion

It is known that during the Chemical Bath process two types of reactions are carried out: One of them produces precipitates of the material. The other one produces film on the substrate. Both of them are in competition and depend on the concentration of the precursors agent. The scope of this work is dealing with the second one.

Furthermore, we assume that the deposition process is performed according MWCBD to model Stranski-Krastanov [10] that is characterized by the growth of layers (2D films) and islands (3D microstructure). The transition layer-by-layer growth to island occurs after a critical thickness of the layer, which is highly dependent on the surface energy and lattice parameters of the substrate and the film. Figure 1 reports the growth measurement of the total thickness (2D and 3D). Where it can be seen that the thickness of the deposit increases as the pH in the solution increases. In this case, NH_4OH is the reactive limiting of the process. When increasing the amount of this reactive, increases the reaction yield, which is reflected on the thickness of the deposit.

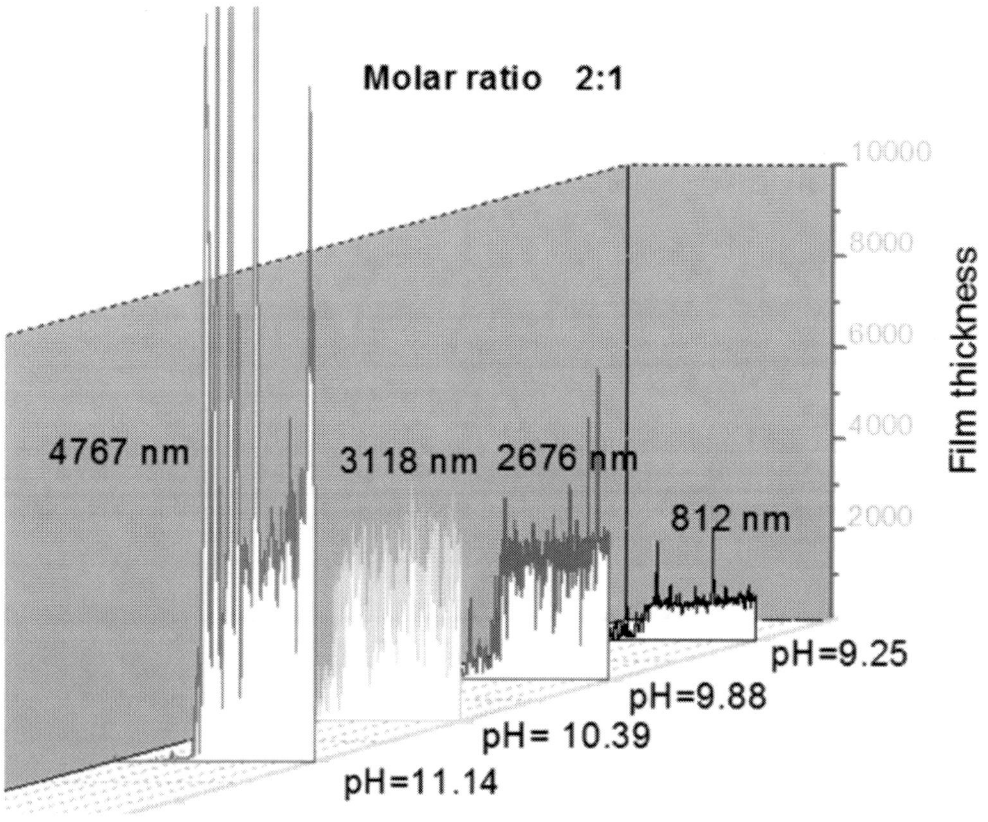

Figure 1. Graph of the ZnO film average thickness on the pH of the precursor solution

XRD patterns from the samples synthesized with solutions of 2:1 molar ratios and different pH values are presented in Figure 2. It shows that ZnO has a polycrystalline hexagonal structure wurtzite type, according to the pattern 36-1451 of the database of PDF-2 of ICDD. The results evidence the pH influence on the nucleation and growth process.

The diffraction pattern of material having a pH = 9.25 (Figure 2) has the peak (0002) more intense, suggesting that preferential growth is present along the direction [0001], the peak observed may be due to a first deposit (2D) highly crystalline.

Through the change in pH, it is possible to control the ratio of the growth rate between Zn^{+2} ions and OH^- ions [11]. The increase of OH^- ions inhibits growth in the plane direction (0001) and induces the growth in the family of planes [1010], [12, 13].

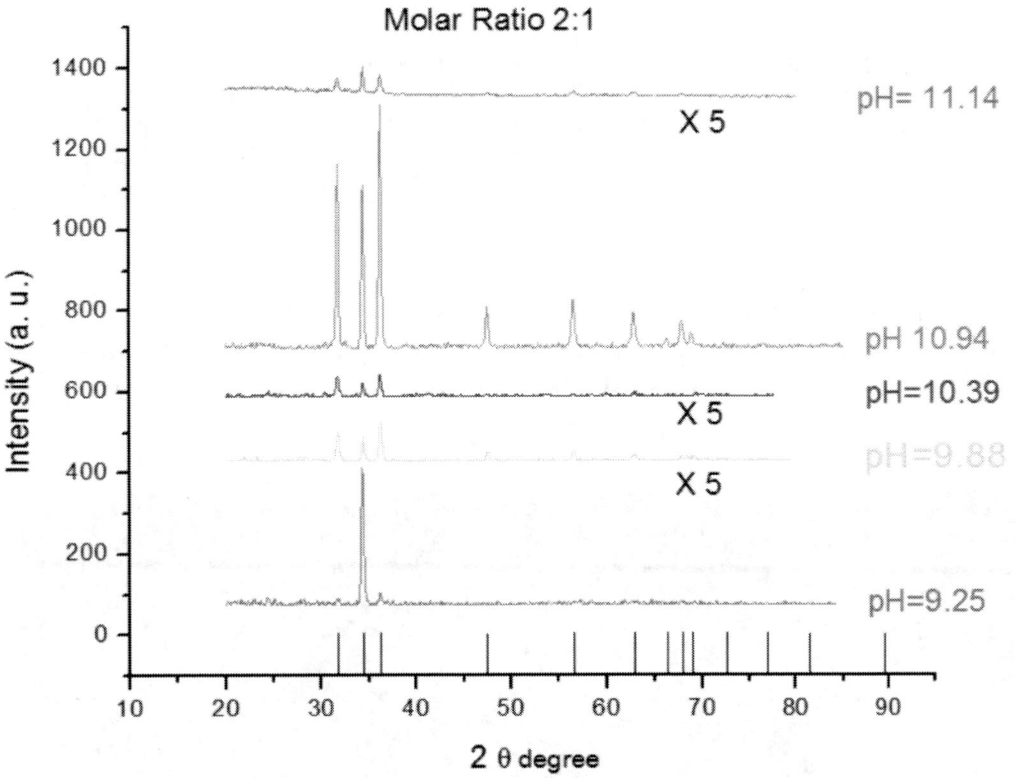

Figure 2. Diffraction patterns of ZnO films obtained from solutions with molar ratio 2:1, with different pH.

pH	Crystal size in different directions			Average crystal size
	(100)	(002)	(101)	
9.25		46.53		46.53
9.88	21.86	22.61	24.005	22.82
10.39	25.45	29.76	27.84	27.68
10.94	25.09	29.42	27.26	27.25

Table 1. Crystal size calculated from the diffraction patterns obtained from samples with different pH.

The average crystal sizes in the main orientations were calculated by using the Scherrer equation; the results are shown in Table 1. Films obtained with pH of 9.25 have a preferential growth in the orientation (002) and a size of 46.53 nm. For polycrystalline films the obtained average crystal size is between 20 and 30 nm, at higher pH.

Atomic Force Microscopy (AFM) was used for to obtain the topography of the deposits (Figure 4). Deposits with larger particle size were obtained at higher concentration of the dissolution [14] in concordance with the results obtained by profilometry, In the Figure 4, it is seen the evolution on the surface of the films in relation to the increase in the pH in the precursor solution.

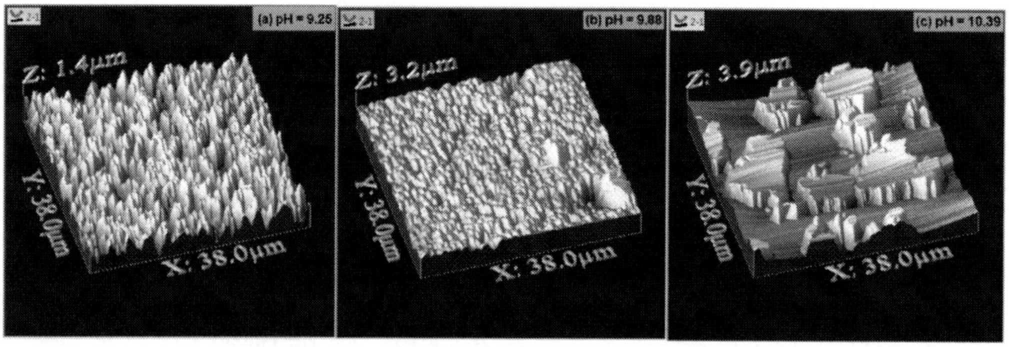

Figure 4. Topography of films obtained from 2:1 solutions: a) pH= 9.25, b) pH=9.88, c) pH=10.39.

4 Conclusions

ZnO films with good quality crystalline were obtained by microwave chemical bath deposition technique, which was demonstrated by XRD characterization, this analysis demonstrated that the films have the hexagonal wurtzite structure and by the Scherrer equation we found that the crystal size of the films decreases by increasing the pH. As a result of increasing pH, the topography of the films and the deposit thickness increases. It is possible to obtain films with a preferential growth orientation, as well as polycrystalline deposits by adjusting the pH in the solution used for its growth.

References

[1] Y. Y. Kim, C. H. Ahn, S. W. Kang, B. H. Kong, S. K. Mohanta, H. K. Cho, J. Y. Le, and H. S. Kim, *J. Matter Sci: Matter Electron* (2008) **19**, 749-754.
[2] Hongbing Lu, Tian Yu, Hu Meifeng, Shuai Min, LI Jinchai, *Wuhan University Journal of Natural Sciences* (2007) **12** No.6.
[3] Cordt Zollfrank, Charles R. Rambo, Miroslaw Batentschuk, and Peter Greil, *J. Matter Sci* (2007) **42** 6325-6330.
[4] WD Yu, X.M. Li, X.D. Gao, P.S. Qiu, W.X. Cheng, A.L. Ding, *Appl. Phys A* **79**, (2004), 453-456.
[5] O. A. Lyapina, A. N. Baranov, G. N. Panin, A. V. Knotko and O. V. Kononenko, *Inorganic Materials*, 2008, **44**, No. 8, p. 846-852. Pleiades Publishing, Ltd., 2008.
[6] *Aqueous Microwave Assisted Chemistry, Synthesis and Catalysis*. Ed. by Vivek Polshettiwar and Rejender S. Varma. Royal Society of Chemistry 2010, p. 189-190.
[7] Chung-Hsin Lu*, Chi-Hsien Yeh, *Ceramics International* **26** (2000) 351-357
[8] D. Vernardou, G. Kenanakis, S. Couris, E. Koudoumas, E. Kymakis, N. Katsarakis, *Thin Solid Films* 515 (2007) 8764–8767
[9] I. Horcas et. al. *Rev. Sci. Instrum.* **78** (2007) 013705.
[10] D. L. Smith, *Thin-Film Deposition: Principles and Practice.* New York: McGraw-Hill, 1995.
[11] Yunyan Zhang and Jin Mu, *Nanotechnology* **18** (2007) 075606 (6pp).
[12] Ya-Jing Zhang, Qi Yao, Ying Zhang, Tie-Yu Cui, Da Li, Wei Liu, Whitmore Lawrence, and Zhi-Dong Zhang, *Crystal Growth & Design, Vol. 8, No. 9, 2008*, 3206-3212.
[13] Huibo Chen, Xiang Wu, Lihong Gong, Cai Ye, Fengyu Qu, Guozhen Shen, *Nanoscale Res Lett* (2010) 5:570–575.
[14] Chung-Hsin Lu, Chi-Hsien Yeh, *Ceramics International* 26 (2000) 351-357.

Acknowledgements

To PROMEP for their financial support PROMEP/103.5/11/6674.

IOP Conf. Series: Materials Science and Engineering **45** (2013) 012016 doi:10.1088/1757-899X/45/1/012016

Low temperature growth of Zn-ZnO microspheres by atomic hydrogen assisted-HFCVD

R López, G García, T Díaz, A Coyopol, E Rosendo, R Galeazzi, H Juárez and M Pacio

Centro de Investigaciones en Dispositivos Semiconductores
Benemérita Universidad Autónoma de Puebla, 14 sur y Av. San Claudio,
Edif. 137, C.P. 72570, Puebla, México. Tel: (222) 229-5500 ext. 7855

Abstract. Using ZnO pellets and molecular hydrogen as solid source and reactant gas respectively, Zn-ZnO microspheres were grown by hot filament chemical vapor deposition in low temperature growth regime. The presence of Zn-ZnO microspheres was confirmed by scanning electron microscopy (SEM) technique. It is suggested that microspheres formed in the gas phase by nucleation of Zn atoms. Energy dispersive spectroscopy (EDS) and X-ray diffraction (XRD) measurements suggest that microspheres were only superficially oxidized by water vapor species.

1. Introduction

Nowadays, various methods are commonly used for preparation of micro- and nanostructured materials because of their potential applications in electronics, optoelectronics, and chemical gas sensing [1-3]. Among them, chemical vapor deposition (CVD) technique with its experimental configurations such as metal-organic CVD, plasma-enhanced CVD, and low-pressure CVD, have been shown to be reliable methods in obtaining thin films, coatings, and recently nanostructures. Nevertheless, there are some CVD configurations such as hot-filament CVD (HFCVD) technique that have not been much explored. This technique shows some interesting characteristics such as low cost and easy scale-up for the growth over substrates with large areas. Zinc oxide (ZnO), an important wide band gap (3.37 eV) semiconductor, has the prospect of wide variety of applications in many fields due to its interesting optical and electronic properties. ZnO nanostructures with various morphologies, including nanowires, nanorods, and nanotubes, have been successfully fabricated by electrochemical methods, evaporation methods, among others. Recently, synthesis of ZnO hollow microspheres has attracted particular interest because of potential applications in many areas [4, 5]. For example, microspheres are suitable for medical applications since they can serve as small containers for encapsulation on drug delivery. In the present work, we report the formation of Zn-ZnO solid microspheres by the HFCVD technique in low temperature growth regime. It was observed that microspheres not directly growth on the substrate surface, which rules out the vapor-liquid-solid growth mechanism.

2. Experimental

Zn-ZnO microspheres were synthesized by the HFCVD technique in low temperature range of 300-600 ^{0}C. The regime of low temperature for ZnO was determined by using the homologous temperature parameter T_h defined as follows:

$$T_h = \frac{T_p}{T_f}$$

where T_p and T_f are the substrate temperature and the melting point of the grown material respectively. The low temperature regime is considered if the T_h parameter is 0.3 and below. By considering that ZnO has a melting point of 1975 ^{0}C, the substrate temperatures were limited to 600 ^{0}C as the maximum value. Commercial ZnO powder (Mallinckrodt) was compressed to obtain ZnO pellets (0.3 g) which served as raw material. Molecular hydrogen (H_2) was employed as reactant gas. Polished p-type (100) oriented silicon wafers (area 4 cm^2) were used as substrates. The ZnO pellet was loaded into the center of the HFCVD reactor (diameter 60 mm, length 350 mm) and under of a tungsten filament. The scheme of the configuration of the filament, the ZnO pellet, and substrate is shown in figure 1(a). Using an AC voltage of 83.4 V, the filament was heated up to 2000 ^{0}C. The filament-substrate distances were fixed at 9.5, 7.5 and 5.5 mm on each experiment to reach during the process, maximum substrate temperatures of 300, 450, and 600 ^{0}C respectively. The heating rate of the substrate is shown in figure 1(b). The ZnO pellet was placed 2 mm away from the filament and kept at a temperature around 1500 ^{0}C. H_2 gas is introduced into reactor during the experiment and then is activated by contact with the hot filament. This generates a quantity of atomic hydrogen (H*) [6] atoms which are able to decompose the ZnO pellet, forming the gas precursors. After the end of the experiment, the HFCVD reactor was naturally cooled down to room temperature. The chemical reactions on the ZnO pellet and the substrate surface are proposed as follows:

$$ZnO_{(s)} + 2H^*{}_{(g)} \rightarrow Zn_{(g)} + H_2O_{(g)} \quad (1)$$

$$ZnO_{(s)} + H_{2(g)} \rightarrow Zn_{(g)} + H_2O_{(g)}$$

$$Zn_{(g)} + H_2O_{(g)} \rightarrow ZnO_{(s)} + H_{2(g)} \quad (2)$$

Therefore, Zn gas and water vapor precursors are quickly produced (equation (1)) and then, they diffuse to substrate surface where chemical reaction takes place (equation (2)). The XRD diffractograms were measured with a Bruker D8 Discover diffractometer using Cu Kα radiation (1.5418 Å).The morphology and the elemental analysis of the products were characterized using a scanning electron microscope Phillips XL-30.

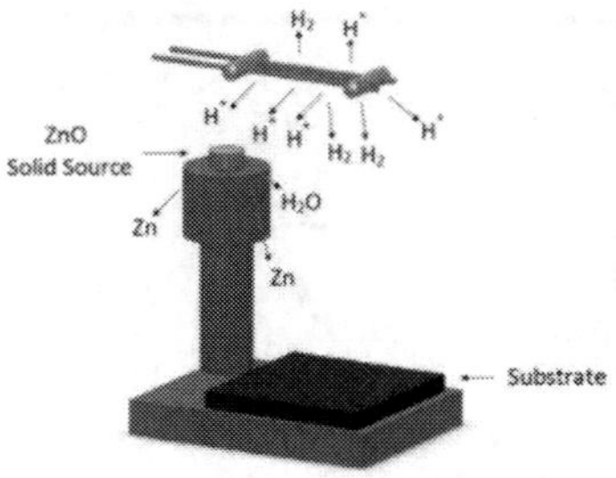

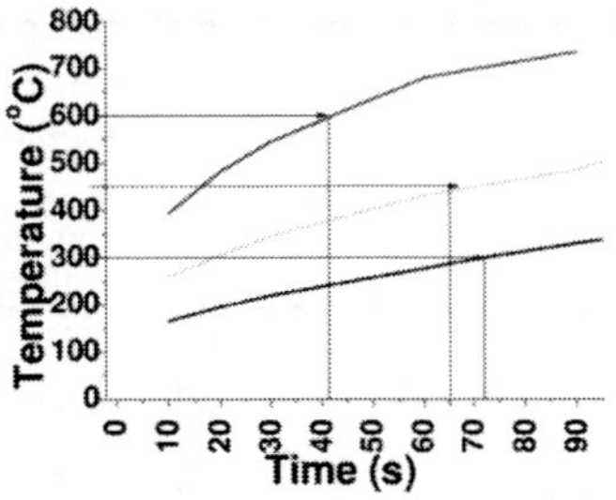

Figure 1. (a) schematic diagram of the Zn-ZnO microspheres growth zone; (b) heating rate of the substrate during deposition process from samples grown by HFCVD technique in low temperature regime.

3. Results and discussion

Figure 2 shows the XRD patterns of the products obtained by HFCVD technique in the temperature range of 300-600 °C. It is observed that a mixture of the Zn (04-0831 JCPDS Card Number) and ZnO (36-1451 JCPDS Card Number) phases in all samples is obtained independently of the substrate temperature. The strongest Zn peak at 2Θ=42.89 ° in the XRD pattern of the sample grown at 300 °C indicates a low contribution of ZnO crystals. The intensity of the Zn peaks decreases with further increase in the substrate temperature, suggesting that Zn crystals are gradually oxidized. The gradual oxidation of Zn has been ascribed to the increase in the surface mobility of Zn and oxygen atoms [7], which induces the nucleation and island growth of ZnO crystals.

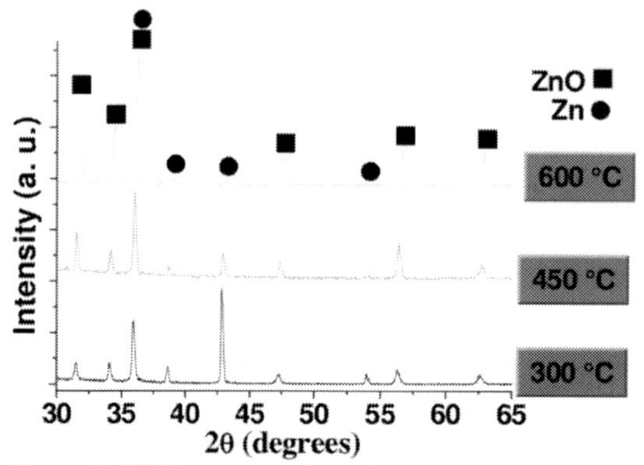

Figure 2. XRD spectra of products grown at 300, 450, and 600 °C by the HFCVD technique.

Figure 3 shows SEM micrographs of samples grown at 300, 450, and 600 °C by the HFCVD technique. It can be seen in Figure 3(a) that a little number of microspheres were formed at 300 °C. The microspheres can be observed on the surface of a layer with fibrous-like morphology, which is quite different from that of silicon substrate. The observation of two morphologies suggests different growth mechanisms. The formation of the fibrous layer is proposed as a result of surface reactions since this layer can be observed at the bottom of the image which shows the top-view of the sample. A similar morphology composed by Zn nanowires was previously observed [8], where the growth of such Zn nanowires was attributed to the so-called vapor-solid process. The presence of Zn and a weak oxygen signal in the EDS results measured from the fibrous layer (not shown here) indicates the formation of a superficially oxidized Zn fibrous layer. Moreover, the formation of Zn microspheres (see inset in Figure 3(a)) suggests that the material was liquid before solidification and that the condensation of these microspheres takes place before they reach the surface of the Zn fibrous layer [9]. Thus, it seems that the Zn microspheres condensed during the natural cooling of the HFCVD reactor as they are observed on the surface of the Zn fibrous layer, which grows during the experiment. Higher growth temperatures lead to an increase in the number of Zn microspheres (Figures 3(b) and 3(c)). EDS spectra (see insets in Figures 3(b) and 3(c)) suggest that the Zn microspheres have been partially oxidized into ZnO at 450 and 600 °C. This could be attributed to the higher number of oxidant species present in the gas phase during the natural cooling of the HFCVD reactor since the substrate surface is closer to the hot filament.

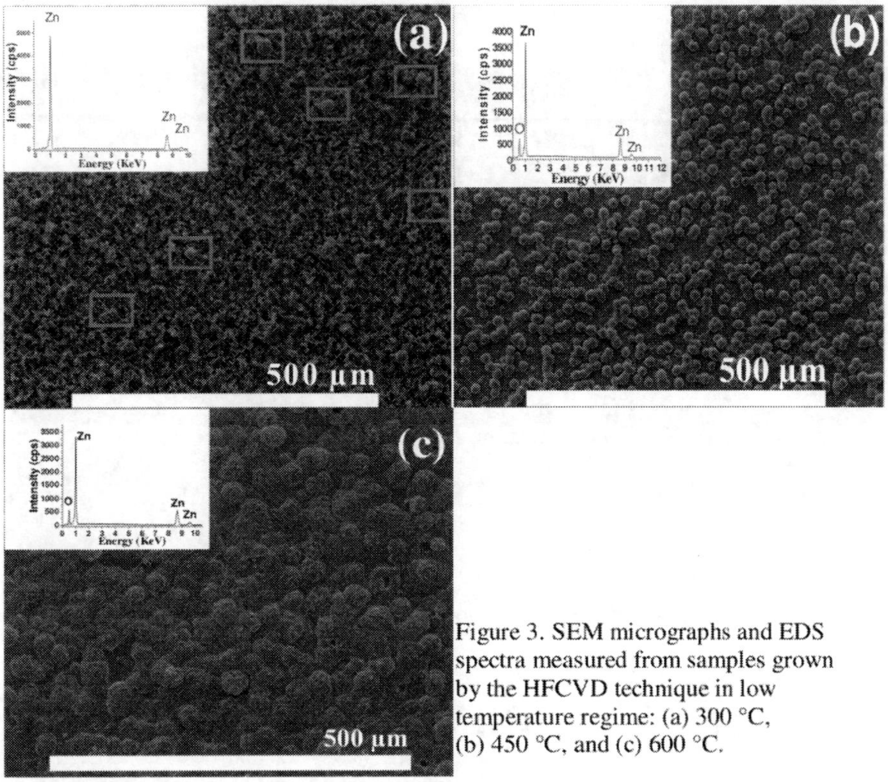

Figure 3. SEM micrographs and EDS spectra measured from samples grown by the HFCVD technique in low temperature regime: (a) 300 °C, (b) 450 °C, and (c) 600 °C.

4. Conclusions

Zn-ZnO microspheres were grown by HFCVD technique. XRD results showed a mixture of Zn and ZnO crystalline phases in all samples. The Zn crystals were gradually oxidized as the substrate temperature was increased. SEM and EDS results showed the formation of Zn microspheres at 300 ^{0}C and Zn-ZnO microspheres at 450 and 600 ^{0}C. We suggested a gas formation and subsequent condensation of Zn microspheres during the natural cooling of the HFCVD reactor as they were observed on the surface of a layer which was formed on the substrate surface during the experiment.

5. References

[1] Kind H, Yan H, Law M, Messer B and Yang P 2002 *Adv. Mater.* **14** 158-160.
[2] Zhang BP, Binh NT, Segawa Y, Wakatsuki K and Usami N 2003 *Appl. Phys. Lett.* **83** 1635-1637.
[3] Du G, Ma Y, Zhang Y and Yang T 2005 *Appl. Phys. Lett.* **87** 213103-213105.
[4] Caruso F 2000 *Chem. Eur. J.* **6** 413-417.
[5] Fowler CE, Khushalan D and Mann S 2001 *J. Mater. Chem.* **11** 1968-1972.
[6] Harris SJ and Weiner AM 1993 *J. Appl. Phys.* **74** 1022-1026.
[7] Ye JD, Gu SL, Zhu SM, Qin F, Hu LQ, Ren L, Zhang R, Shi Y and Zheng YD 2004 *Appl. Phys. A* **78** 761-764.
[8] Kar S, Goshal T and Chaudhuri S 2006 *Chem. Phys. Lett.* **419** 2981-2996.
[9] Weidenkaff A, Reller A, Sibieude F, Wokaun A and Steinfeld A 2000 *Chem. Mater.* **12** 2175-2181.

IOP Conf. Series: Materials Science and Engineering **45** (2013) 012017 doi:10.1088/1757-899X/45/1/012017

Recovery evaluation of organophosphorus pesticides from bee pollen by matrix solid-phase dispersion extraction using sorbents based on silica and titania

C Torres-Perea[1] D Muñoz-Rodríguez[2] S Medina-Peralta[3] C Carrera-Figueiras[2] and Y B Moguel-Ordóñez[4]

[1] Facultad Ciencias Químicas, Universidad Veracruzana, Orizaba, Veracruz, México.

[2] Facultad de Ingeniería Química, Campus de Ingenierías y Ciencias Exactas, Universidad Autónoma de Yucatán, Mérida, Yucatán, México.

[3] Facultad de Matemáticas, Campus de Ingenierías y Ciencias Exactas, Universidad Autónoma de Yucatán, Mérida, Yucatán, México.

[4] Instituto Nacional de Investigaciones Forestales, Agrícolas y Pecuarias, CE Mocochá, Mocochá, Yucatán, México

E-mail: david.mr@uady.mx

Abstract. This work focused on the evaluation of the recovery of organophosphorus pesticides from bee pollen after matrix solid phase-dispersion extraction (MSPD). Materials based on silica, titania and titania modified with polivylnylimidazole or polyestirene were used as adsorbents for the extraction of pesticides. Small amounts of fortified pollen (0.1 g, at 1 micro-g/g of pesticides), adsorbent (0.4 g) and solvent elution (1 mL de acetonitrile - ACN) were used in the extractions. For recovery evaluation, pollen extracts were analyzed by gas chromatography coupled with mass spectrometry.

1. Introduction

Matrix solid-phase dispersion (MSPD) is a sample preparation technique applied widely to solid, semisolid or viscous samples [1]. It has been employed for the extraction of organic contaminants [2] from many different samples such as propolis [3], pollen [4] and honey bees [5]. The steps for MSPD involve mixing the sample in a glass mortar in the presence of a solid support (adsorbent), packing the mixture in a glass (or polypropylene column) and eluting the analytes from the column with an appropriate solvent [1]. In the last step, interfering compounds are retained selectively on the solid phase. An alternative is the elution of interferences with a suitable solvent (wash step) followed by the elution of analytes retained in the MSPD column with another solvent [6].

The factors that affect the efficiency of extractions by MSPD are (1) the physicochemical properties of both the solid support and the bonded phase, (2) nature of matrix sample, (3) matrix modification, (4) solvent elution and its sequence addition [1]. Among these factors, the former affects the retention/elution of the analytes and the dispersion of the solid matrix sample onto the solid support. Thus, the solid supports typically reported in MSPD are commercial. These are based in underivatized silica or derivatized with organic groups such as octadecyl (-C18), cyanopropyl (-$(CH_2)_3CN$) and aminopropyl (-$(CH_2)_3NH_2$) [7]. However, even though materials-science provides the tools for the development of new solid supports and organic phases, there are few reports on the development and application of "home-made" solid supports [8]. Thus, titania (TiO_2) organic-inorganic hybrid materials have not been employed as the solid phase in MSPD. In comparison with silica based materials, the main advantages of TiO_2 materials are its stability at pH 1-14 and its mechanical strength.

The aim of this work was the evaluation of new solid supports for the extraction of organophosphorous pesticides from bee pollen. One of the solid supports tested was commercially available and is based on silica (Florisil). The others were synthesized in the laboratory and are based on titania (TiO_2) and titania modified with polyvilnylimidazole (TiO_2-PVI) or polystyrene (TiO_2-PST). The bee pollen was selected as sample due to its complex chemical composition [9] and pesticides (dichlorvos, diazinon, methyl parathion, malathion and coumaphos) because of its wide use against pests in crop fields and bee hives.

2. Experimental

2.1. Chemicals and materials

Florisil was obtained from Sigma-Aldrich (St. Louis, MO, USA). Analytical grade anhydrous sodium sulfate (Na_2SO_4) and solvents (acetonitrile, isooctane, ethyl acetate) were purchased from J. T. Baker (Mallinckrodt Baker, Phillipsburg, NJ, USA). Pesticide standards (at least 97% pure) [dichlorvos (DCV), diazinon (DZN), methyl parathion (MPT), malathion (MLT) and coumaphos (CMF)] were obtained from ChemService (West Chester, PA, USA). Individual stock solutions were prepared in ethyl acetate and stored at -18°C. Working standard solutions were prepared by diluting the stock solutions with ethyl acetate.

2.2. Fortified bee pollen samples

Bee pollen was spiked by the addition of pesticide standard solution in ethyl acetate (100 µL at 1 µg/mL) to 0.1 g of pollen to obtain a concentration of 1 µg/g. Then, the sample was left to stand (60 min) to allow pesticides to incorporate into the pollen.

2.3. Experimental design

In this stage of methodology, Florisil was used as solid support in MSPD. An unreplicated 2^4 factorial design [10] was used to study the experimental factors that affect the efficiency of MSPD extraction: (1) mass of solid support (A: 0.2 g, 0.4 g), (2) conditioning solid support (B: no, yes), chemical nature of elution solvent (C: ethyl acetate, acetonitrile) and solvent volume (D: 1 mL, 2 mL). The recovery percent (%R) of pesticides in pollen extracts was the variable response.

2.4. MSPD extraction

Bee pollen (0.1g) and solid support (Florisil, TiO_2-PVI, TiO_2-PST or TiO_2) were placed in a mortar. Then, both were dispersed with a pestle to obtain a homogeneous mixture which was packed in a polypropylene column with a filter paper disc at the bottom. Next, it was compressed softly to eliminate air pockets. The pesticides were eluted with organic solvent which was evaporated to dryness with a gentle air flow. The extract was reconstituted with isooctane and Na_2SO_4 was added (0.2 g). The extract was frozen (-3°C, 60 min) and centrifuged (50 min, -5°C) to precipitate high molecular weight compounds. Finally, the supernatant was analyzed by GC/MSD.

2.5. Instrumentation and chromatographic conditions

An Agilent Technologies 6890N gas chromatograph coupled with a mass spectrometer 5973N (GC/MSD) was used. A fused-silica column Equity-5 (5% phenyl-95% polydimethylsiloxane; 30 m × 0.25 mm ID, 0.25 µm), supplied by SUPELCO (Bellefonte, PA, USA) was employed with He (purity

99.999) as carrier gas (flow rate: 1.0 mL·min^{-1}). The oven temperature was programmed as follows: 120°C for 3 min directly to 280°C at 20°C·min^{-1} and holding for 5 min (solvent delay: 3 min). The inlet temperature was 250°C and 1 μL of the extract was injected in a splitless mode (1.5 min). The eluent from the GC column was transferred into the MSD via a transfer line (at 280°C). Typical conditions were optimized through the autotune software option. The electron impact mode (70eV) was used as ionization source (250°C) and masses were monitored between 50-400 m/z (SCAN mode). The temperature of the quadrupole was 150°C. The analysis was performed in selected ion monitoring (SIM). The mass-charge ratios (m/z) for the ions for each pesticide were 109, 185, 220 for DCV; 179, 199, 304 for DZN; 109, 125, 263 for MPT; 125, 158, 173 for MLT and 226, 334 and 362 for CMF.

2.6. Efficiency of the MSPD extraction with SiO_2-poly (vinyl-imidazole)

Recovery experiments were done at three fortified levels (0.1, 0.5 y 1.0 μg/g) to evaluate the efficiency of the solid supports in MSPD. Recoveries were calculated through external calibration and pesticide standards were prepared by dilution with bee pollen extracts to minimize matrix effect.

3. Results and discussion

Recoveries of 1-110% were obtained by MSPD using Florisil as the solid support. The lowest recovery was for DCV and the highest for DZN. Recoveries higher than 120% were obtained for MPT, MLT and CMF in experiments where ethyl acetate (1 mL) was used as the eluent. This is ascribed to the elution of interfering compounds having the same retention time of MPT, MLT and CMF. It is reported that ethyl acetate elutes large amounts of non-polar interfering compounds [11]. In contrast, lowest recoveries (<60%) were obtained with acetonitrile (2 mL) since we believe that the analytes may evaporate during the increased time required to evaporate the large volume of solvent.

The experimental conditions that enabled good recovery percent for all of the pesticides studied were as follows: 0.4 g of Florisil without conditioning and 1 mL of acetonitrile. These conditions were applied for the extraction of fortified bee pollen at 0.1, 0.5 and 1.0 μg/g of fortification. At the lowest level, DCV and MPT were not detected and recoveries for MLT were lower than 11%. However, DZN showed a good recovery (105%). At 0.5 and 1.0 μg/g of fortification, recoveries for DCV were 69% and 48%, respectively. Nevertheless, recoveries for DZN, MPT and MLT ranged between 76 to 114%.

We evaluate the efficiency of TiO_2-PVI, TiO_2-PST or TiO_2 using 0.4 g of those materials as solid support instead of Florisil and 1.0 mL of acetonitrile. Recoveries are presented in table 1.

Table 1. Pesticide recoveries (%R) using silica and titania based materials as the solid support and spiked pollen at 1.0 μg g^{-1} (n=3).

Pesticide	Material	%R	% RSD
Dichlorvos	Florisil	48	5
	TiO_2-PVI	33	36
	TiO_2-PST	32	10
	TiO_2	22	11
Diazinon	Florisil	96	4
	TiO_2-PVI	79	8
	TiO_2-PST	73	8
	TiO_2	86	22
Malathion	Florisil	77	14
	TiO_2-PVI	84	49
	TiO_2-PST	89	15
	TiO_2	99	16

Two pesticides (MPT and CMF) were not detected in chromatograms probably due to its retention into the TiO_2 materials. For DZN and MLT, recoveries were higher than 70% with the three materials based in TiO_2. However, recoveries obtained with TiO_2 were relatively higher than those obtained with TiO_2-PVI or TiO_2-PST. The lowest recoveries were for DCV (%R <35).

The recoveries for DZN and MLT with TiO_2 materials were similar to those achieved with Florisil. DCV recoveries obtained with both kinds of material were lower than 50%. This pesticide has a high vapor pressure and it can be lost during the solvent evaporation step [12].

For the three pesticides extracted with TiO_2-PST the precision (%RSD) was relatively higher than those obtained with Florisil, but it was equal to or lower than 15% as required for pesticide residue analysis [13]. Finally, precision of recoveries for MLT was close to %RSD obtained with Florisil. %RSD values of DCV, DZN and MLT were 20% higher than those obtained with TiO_2 or TiO_2-PVI.

4. Conclusions

Solid phases based on TiO_2 were used in MSPD to extract organophosphorous pesticides from spike bee pollen and pesticide recoveries were determinated.

The experimental conditions that yield pesticide recoveries from 70 to 120% were 0.4 g of Florisil (no conditioning) and acetonitrile (1 mL). Recoveries obtained for DZN, MPT and MLT with Florisil in spiked bee pollen (0.5 and 1.0 µg/g) were 76 to 114%. DCV presented recoveries from 47 to 69%. Low recoveries (<11%) were obtained at 0.1 µg/g or signals were not detected for pesticides.

With titania materials, DCV, DZN and MLT were detected in spiked pollen extracts. The recoveries for the first pesticide were lower than 35%. The last two pesticides presented recoveries higher than 70%. MPT and CMF were not eluted from the MSPD column.

Recoveries for DZN and MLT with the three kind of TiO_2 materials were similar to those obtained with Florisil. Precision of pesticide recoveries was acceptable (RSD≤15%) with TiO_2-PST.

5. Acknowledgements

The authors thank the Consejo Nacional de Ciencia y Tecnología (CONACyT) for financial support to carry out this work through the basic science project 83390. The authors thank Mr. Daniel Joseph Gleber for manuscript revision.

References

1. Barker SA (2007) Journal of Biochemical and Biophysical Methods 70: 151-162
2. Bogialli S, Corcia Di A (2007) Journal of Biochemical and Biophysical Methods 70:163 - 179
3. Santana T, Aquino A, Silveira H, Navickiene S (2008) Anal bioanal chem 390:1425-1430
4. Jiménez JJ, Bernal JL, Nozal del MJ, Martín MT, Mayo R (2007) J Chromatogr A 1146:8-16.
5. Totti S, Fernández M, Ghini S, Picó Y, Fini F, Mañes J, Girotti S (2006) Talanta 69:724 - 729
6. Barker S (2000) Journal of Chromatography A 885: 115-127
7. Muñoz-Rodríguez D, Ávila-Ortega A, Barrón-Zambrano JA, Carrera-Figueiras C (2011) INVURNUS 6:22-30
8. Capriotti AL, Cavaliere C, Giansanti P, Gubbiotti R, Samperi R, Lagana A (2010) J Chromatogr A 1217:2521-2532
9. Campos MGR, Bogdanov S, de Almeida-Muradian LB, Szczesna T, Mancebo Y, Frigerio C, Ferreira F (2008) J Apic Res and Bee World 47:156–163
10. Montgomery DC (2004) Diseño y análisis de experimentos. 2a edición edn. Limusa Wiley, México, D.F.
11. Anastassiades M, Lehotay SJ, Štajnbaher D, Schenck FJ (2003) J of AOAC Internacional 86:412-431
12. Lehotay SJ, K M (2005) Journal of AOAC International 88:631 - 638
13. Hill ARC, Reynolds SL (1999) Analyst 124:953–958

IOP Conf. Series: Materials Science and Engineering **45** (2013) 012018 doi:10.1088/1757-899X/45/1/012018

Adsorption of cationic dye on a biohybrid SiO_2-alginate

J A Barrón Zambrano[1], A J López-Pérez[2], A Ávila Ortega[1], D Muñoz Rodríguez[1] y C Carrera Figueiras[1]

[1]Universidad Autónoma de Yucatán, Campus de Ciencias Exactas e Ingenierías, Facultad de Ingeniería Química.
[2]Universidad Veracruzana, Facultad de Ciencias Químicas

E-mail: jesus.barron@uady.mx

Abstract. In this work, a biohybrid material based on SiO_2-alginate was obtained and its properties of adsorption evaluated using methylene blue as a model dye. The experimental results showed that the biohybrid SiO_2-alginate has a higher adsorption ability compared to their base compounds (SiO_2 and alginate). Methylene blue adsorption is pH dependent, resulting in a maximum adsorption at pH = 8. The sorption kinetics rate is similar to SiO_2. Kinetic data were fitted to a model of pseudosecond order. The experimental isotherms fit well the Langmuir model.

1. Introduction

With the rapid development of large industries and the increasing growth of urban areas, the discharge of effluents contaminated with dyes represents a serious risk to health and the environment. Various technologies have been developed and tested in colored industrial effluent treatment systems. In this reference framework, the activated carbon adsorption has proven to be the most promising strategy, but its high cost restricts its large scale implementation. Hence, the exploration of alternative materials with similar properties, natural origin, easy acquisition, minimal economic value and environmentally friendly is imperative. In recent years there has been a growing interest in the development of new hybrid materials and their potential application as adsorbents for the removal of different types of pollutants [1, 2]

Researchers have reported studies on compounds derived from silica, which have been successfully applied as adsorbents in the removal of pollutants such as aromatic compounds and some heavy metals. An et al. [3, 4] developed a new adsorbent material PAM/SiO_2 for removal of 2,4,6-trinitrotoluene (TNT) and aniline. The material produced had a strong adsorption ability and reusability for the removal of both, TNT and aniline.

On the other hand, chitosan-silica hybrid materials obtained as microparticles, have been prepared by the sol-gel method using TEOS or oligomers in the presence of the polymer polyethoxysiloxanes. These materials may be used as stationary phase in HPLC [5]. The hybrid particles, based on algal polysaccharides such as alginate and carrageenan, appear to be a good candidate for the formation of silica-based biocomposites. Especially, alginate is a copolymer of L-guluronic and D-mannuronic acid. It can be reversibly gelled by the addition of divalent metal, like calcium [6, 7].

The biopolymer/silica composites are suitable for the design of membranes and coatings used in drug detection systems [6], as sorbent materials for some metals such as copper [7], and for the encapsulation of bioactive molecules such as enzymes, antibodies, yeast and plant cells and even bacteria, resulting in biomaterials for various biotechnological applications [8]. In these examples, the interaction of the biological and inorganic components has synergistic effects that lead to hybrid materials with improved mechanical strength, higher thermal and chemical stability and, in some cases, functional properties.

In this paper, a biohybrid material based on SiO_2-alginate was obtained and its properties of adsorption evaluated using a cationic dye, namely Methylene Blue (MB) as a model dye. The effects of pH, contact time and initial dye concentration on the dye removal were investigated. The kinetic parameters were also calculated to determine rate constants. The equilibrium experimental data were fitted to the Langmuir model.

2. Materials and Methods

2.1. Synthesis of adsorbents

SiO_2 was synthesized by mixing Tetraethyl Orthosilicate (TEOS), ethanol and distilled water in a beaker with slow stirring. 1 M HCl was added to the mixture until pH 2 and allowed to stir for 30 minutes. 0.1 M NaOH was added until pH 7 and then continued with stirring to obtain a gel. The gel obtained was allowed to grow for 24 h, and then washed with water. Subsequently, the product was dried at 85 °C in air oven to constant weight.

Calcium alginate powders were obtained by spraying 600 mL of 2% sodium alginate solution on 500 mL of 0.05 M $CaCl_2$ solution. The formed calcium alginate particles were allowed to mature for 24 h. After that, they were filtered, washed with distilled water and dried at 50 °C in air oven for 24 h. After drying, the particles were maintained in a desiccator.

Biohybrid sorbent was synthesized by mixing in a beaker with slow stirring, TEOS, ethanol and distilled water. The ratio TEOS/H_2O was kept constant and equal to 1. 1 M HCl was added to the mixture until pH 2 and allowed to stir for 30 minutes. The above mixture was added dropwise to 500 mL of 2% sodium alginate solution of under rapid stirring until total homogenization of the two solutions. After complete addition, agitation was reduced and continued to obtain a gel. Subsequently, the gel was dried in vacuum oven at 85 °C and washed with distilled water, then dried again at 85 °C in air oven to constant weight.

2.2. pH Influence

0.1 g of sorbent were put in contact with 30 mL of 50 mg/L initial dye concentration solution, at different pH values (1 to 11) and agitated for 24 h. The resultant solutions were filtered before analyses. Initial and equilibrium dye concentrations were determined by spectrophotometry using UV-VIS spectrophotometer (Thermo Scientific, LR 162800). Absorbance measurements were made at the maximum wavelength of MB (λ= 668 nm). The adsorbed amount of MB at any time, q_e (mg/g), was calculated from mass balance using equation (1):

$$q_e = \frac{(C_0 - C_e)V}{m} \tag{1}$$

where C_0 and C_e are the initial and equilibrium dye concentrations (mg/L), respectively, V is the volume of dye solution (L) and m is the mass (g) of the sorbent used.

2.3. Sorption kinetics

150 mL of MB dye solution (C_o= 50 mg/L) at the pH of maximum adsorption, which was previously determined was put in contact with 0.5 g of sorbent and placed on a mechanical shaker for 10 h. Samples were taken at predetermined time intervals and analyzed as described before. Kinetic data were analyzed with a pseudo-second-order model (equation (2)) [9].

$$\frac{t}{q_t} = \frac{1}{K_2 q_e^2} + \frac{1}{q_e} t \tag{2}$$

where q_e is the adsorption capacity at equilibrium (mg g^{-1}), q_t is the adsorbed amount of sorbate at time t (mg g^{-1}), and K_2 is the pseudo-second order rate constant (g mg^{-1}min^{-1}).

2.4. *Adsorption experiments*

0.1 g of sorbent was added to a bottle containing 30 mL of BM aqueous solution of desired concentrations (10-100 mg/L) at the pH of maximum adsorption previously determined. The bottles were subsequently capped and placed on a mechanical shaker at a speed of 200 rpm at (20±2) °C for 3 h. Experimental data were fitted to the isotherm model of Langmuir (equation (3)) [10].

$$q_e = \frac{q_{max}K_L C_e}{1+K_L C_e} \tag{3}$$

where C_e is the supernatant concentration at the equilibrium (mg L^{-1}), K_L is the Langmuir affinity constant (L mg^{-1}), and q_{max} is the maximum adsorption capacity of the sorbent (mg g^{-1}) assuming a monolayer coverage of the sorbent.

3. Results and discussions

3.1. *pH Influence*

Figure 1 shows the influence of initial pH on the adsorption of MB onto SiO_2/alginate biohybrid, SiO_2 and alginate. It can be seen that in the case of the basis compounds the adsorption is pH dependent, however, this effect is less important in the case of the biohybrid, probably as a result of a synergic effect of SiO_2 and alginate. This phenomenon allows the use of biohybrid in a wider range of pH conditions with good yields. The maximum adsorption capacity was obtained a pH = 8.

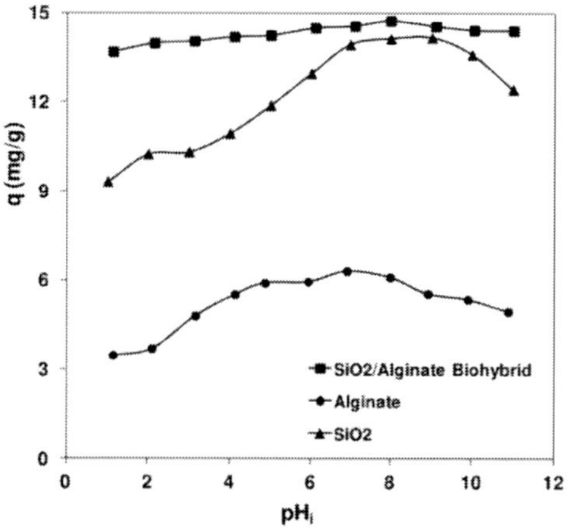

Figure 1. Influence of pH on the sorption of MB on SiO_2, SiO_2/Alginate biohybrid, and alginate.

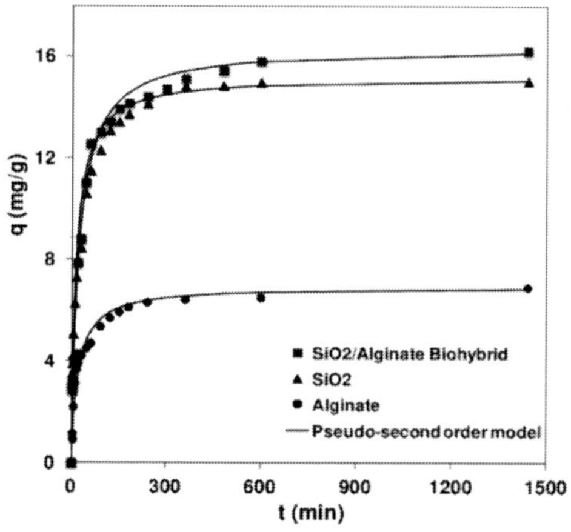

Figure 2. Sorption kinetics of MB on SiO_2, SiO_2/Alginate biohybrid and alginate.

3.2. *Sorption kinetics*

Sorption kinetics for the three sorbents studied is depicted in figure 2. This figure shows that the biohybrid has sorption kinetics similar to SiO_2, but with a slightly better sorption equilibrium capacity. Experimental data fit well to the pseudo-second order model (R^2 > 0.99). The values of K_2 for the biohybrid, SiO_2 and alginate were 0.0027, 0.0045 and 0.0067 respectively, which indicate that the sorbents have similar kinetic sorption rates.

IOP Conf. Series: Materials Science and Engineering **45** (2013) 012018 doi:10.1088/1757-899X/45/1/012018

3.3. Sorption isotherms

Figure 3 shows the sorption isotherms for SiO_2/Alginate biohybrid, SiO_2 and alginate. Experimental data and the parameters of the Langmuir equation show that the biohybrid has a far better maximum adsorption capacity (241.1 mg/ g) than SiO_2 (33.5 mg/g). The experimental sorption isotherm of calcium alginate has an unusual form and could not be modeled with the Langmuir equation.

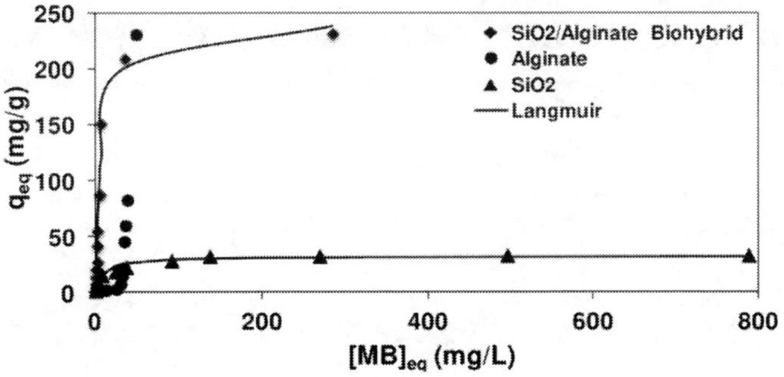

Figure 3. Sorption isotherms of MB on SiO_2/Alginate biohybrid, SiO_2 and alginate

4. Conclusions

The dye sorption capacity of a new SiO_2/alginate biohybrid was evaluated using methylene blue as model dye. Results show that the biohybrid is less affected by the pH than his basis compounds. Maximum dye adsorption capacity for the biohydrid was obtained a pH = 8. Sorption kinetics shows that the biohybrid has similar sorption rate than SiO_2 and alginate. The sorption isotherms demonstrate that the new biohybrid has a better sorption capacity than his basis compounds. Kinetic data were well described by the pseudo-second order kinetic model, while equilibrium data fitted well to the Langmuir model. The biohybrid SiO_2-alginate proved to have excellent adsorptive properties and can be an alternative for the treatment of water contaminated with dyes.

5. Acknowledgements

The authors thank the Consejo Nacional de Ciencia y Tecnología (CONACyT) for financial support to carry out this work through the basic science project 83390

References

[1] Cestari AR, Vieira EFS, Vieira GS,. da Costa LP, Tavares AMG, Loh W and Airoldi C 2009 *J. Hazard. Mater.* **161** 307;
[2] Anbia M, Hariri SA and Ashrazadeh SN 2010 *Appl. Surf. Sci.* **256** 3228
[3] An F, Feng X and Gao B 2009 *J. Hazard. Mater.* **168** 352; An F, Feng X, and Gao B 2009 *Chem. Eng. J.* **151** 183;
[4] An F, Gao B and Feng X 2009 *Chem. Eng. J.* **153** 108
[5] Rashidova S, Shakarova D, Ruzimuradov O, Satubaldieva D, Zalyalieva S, Shpigun O 2004 J. Chromatogr. **800** 49
[6] Xu B, Bartley J P and Johnson, R A 2003 *Appl. Polym. Sci.* **90** 747
[7] Veglio F, Esposito A and Reverberi A 2002 *Hydrometallurgy* **65** 43
[8] Coradin T, Allouche J, Boissière M and Livage J 2006 *Curr. Nanosci.* **2** 219
[9] Ho YS, McKay G 1998 *Chem. Eng. J.* **70**, 115-124
[10] Wong YC, Szeto YS, Cheung WH, McKay G 2003 *Langmuir* **19**, 7888-7894

IOP Conf. Series: Materials Science and Engineering **45** (2013) 012019 doi:10.1088/1757-899X/45/1/012019

A high absorbance material for solar collectors' applications

A I Oliva[1], R D Maldonado[1, 2], E A Díaz[3], A I Montalvo[3]
[1] Centro de Investigación y de Estudios Avanzados del IPN Unidad Mérida, Departamento de Física Aplicada, AP-73-Cordemex, Mérida Yucatán México

[2] Universidad Anáhuac-Mayab, Facultad de Ingeniería. Carretera Mérida-Progreso Km. 15.5, A. P. 96-Cordemex, C. P. 97310, Mérida, Yucatán, México.

[3] DM Energía Verde S A de C V, Calle 32 B No. 402, Chuburná Inn, C. P. 97205 Mérida, Yucatán, México

E-mail: oliva@mda.cinvestav.mx

Abstract. In this work, we proposed a low cost material to be used as an excellent absorber for solar collectors, to increase its thermal efficiency by the high capacity to absorb solar radiation. The material, known as "smoke black" (soot) can be obtained by the incomplete combustion of organic materials, such as the oxygen-acetylene, paraffin, or candles. A comparative analysis between the optical properties (reflectance, absorbance, and emissivity) measured on three covered copper surfaces (without paint, with a commercial matte black paint, and with smoke black) shows amazing optical results for the smoke black. Reflectance values of the smoke black applied over copper surfaces improves 56 times the values obtained from commercial black paints. High values of emissivity (E=0.9988) were measured on the surface covered with smoke black by spectrophotometry in the UV-VIS range, which represents about 7% of increment as compared with the value obtained for commercial black paints (E=0.938). The proposed high absorbance material can be easily applied on any kind of surfaces at low cost.

1. Introduction

Solar collectors for heating fluids require increasing the reception area of the solar radiation, and/or to increase the absorbance of the surface coating in order to improve the thermal efficiency. In a preliminary work [1], a proposal for increasing the reception area of solar radiation without changes in the diameter and number of copper tubes, consists in changing the typical circular cross section for an elliptical geometry with the same diameter of the tubes. Thus, an increment of about 42% in area for solar reception is possible if an eccentricity of ε=0.95 on the elliptical tube is used. Metal makes a good thermal conductor, especially copper and aluminium. In high performance collectors, a "selective surface" is used in which the collector surface is coated with a material having properties of high-absorption and low-emissivity. The selective surface reduces heat-loss caused by infrared radiant emission from the collector to ambient.There exist in the literature several works reporting the good qualities of the selective absorber surface. For example, Peter *et. al.* [2] reported a surface of 99% of absorbance. The studied structure is composed of a reflective metal layer, a roughly quarter-wave layer of lossy dielectric, and a top metal layer that is patterned with an array of subwavelength apertures.

Some companies [3, 4] produce commercial selective solar absorptive layers both for flat plate and tubular collector solar systems but with a major cost. The layer configuration is formed by three parts: an IR-reflective part to guarantee low emissivity, an oxide solar absorptive part and an antireflective part for maximum solar absorption, as well as to resist degradation. Recently, a new type of ultra-thin meta-materials [5] was proposed to obtain broadband absorbers through surface plasmons.

2. Theory

When the radiative energy arrives on a surface, the total energy can be divided in three fractions: absorbed (A), reflective (R) and transmitted (T). Thus, by energy conservation, A+R+T= 1. On the other hand, according to Kirchoffs law, an isothermal body in thermodynamic equilibrium possess emissivity equivalent to its absorbance. An ideal black body assume that the absorbed energy is totally emitted, such that its emissivity is E=1. If the body is opaque to the visible wavelength range (T=0), as our case, then, energy conservation reduces to A+R=1. With these considerations, the reflectance R can be determined by measuring the absorbance A, and consequently, the emissivity E of the black body.

3. Experimental Setup

Samples preparation

Groups of three samples of laminated cooper of 25 mm x 25 mm as obtained from commercial tubes of 1.27 cm-diameter used for plane solar collectors were prepared. Before preparation, samples were ultrasonically cleaned with acetone, and isopropyl alcohol. The first group of samples were analyzed without coating on the copper sample. The second group of samples was totally covered with commercial mate black paint with an aerosol-type application. Third group of samples was completely covered with "smoke black" as obtained from the incomplete combustion provided by an oxygen-acetylene welder equipment. For the third group, we suppose that the covered surface is mainly formed by diminutive particles of carbon as a result of the incomplete combustion and maybe some disordered nanotubes can be included.

Spectrophotometry measurements

Optical measurements (absorbance and emissivity) on samples were done with a Jobin Yvon/Spex UV-Vis spectrophotometer in the 300-800 nm wavelength range. The light source used was an incandescent lamp (white light) with a solid state germanium detector. For reflectance measurements a home made sample holder was used which is capable to measure the incident and reflected light at 45°. The experimental setup used is similar to the proposed by the National Centre of Metrology in México [6].

4. Results

Figure 1 shows the results of the reflectance of the three samples as a function of the wavelength. The vertical scale is logarithmic for better appreciation. Arrows shows the values of the reflectance for 400 nm and 800 nm, respectively. From Figure 1, it is observed the reflectance reduction when the copper surface is covered with a black paint. Minor reflectance is observed for the surface with a smoke black cover. The reduced value of the surface (R=0.05%) covered with smoke black seems as a black body.

Figure 2 shows the emissivity (absorbance) as measured for the two samples of copper as a function of the wavelength. In figure 2 is shown the copper samples covered with black paint and without covering. Arrows indicate the values of the emissivity (absorbance) at 400 and 800 nm. Note the important advantage when the black paint in included on copper as compared with the surface copper only.

The copper sample covered with smoke black is shown in Figure 3 alone given the larger differences in the absorbance scale. From Figure 3 it is observed that the smoke black cover presents almost ideal black body behaviour. The high absorbance measured on the smoke black (A=99.95%)

makes it a good candidate to be used for recovering the metallic surface of plane solar collectors for increasing its absorbance. A higher value of emissivity (absorbance) was also reported on surfaces prepared with vertical carbon nanotubes [7].

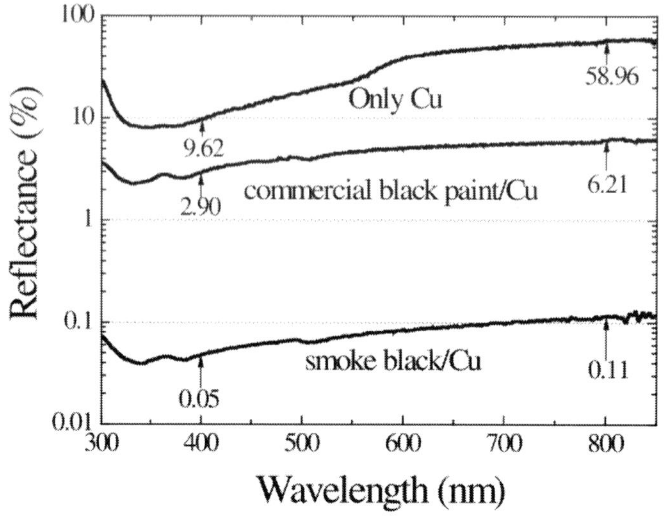

Figure 1. Optical reflectance as measured from three different copper surfaces.

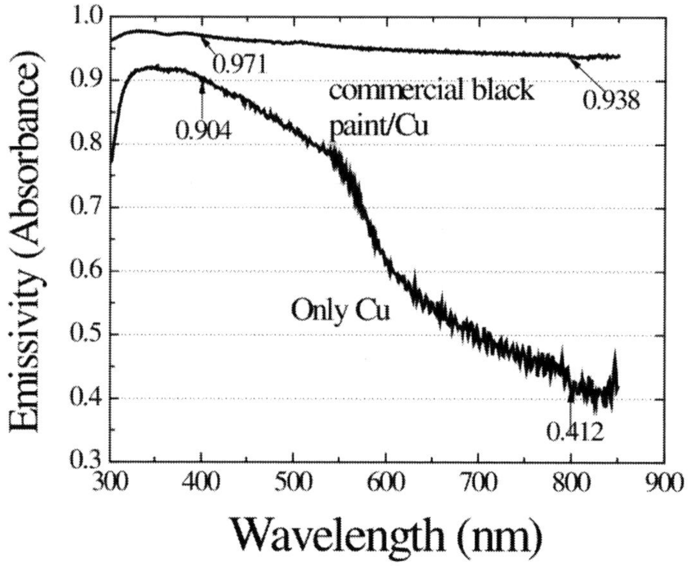

Figure 2. Emissivity (absorbance) of a copper surface without and with commercial mate black paint.

5. Conclusions

Three surfaces of copper for solar collector applications were optically analyzed in order to determine the emissivity (absorbance) value. Copper surfaces requires to be covered with some kind of black paint in order to increase its absorbance; otherwise, the reflective fraction increases, and diminishing the thermal efficiency. However, smoke black as obtained by the incomplete combustion of a oxi-acetylene welder equipment and applied on copper surface, demonstrate to have higher absorbance than other commercial black paints. A near black body behaviour was measured on the copper surface when ultra-low reflectance of 99.95% was measured for smoke black paint. The smoke black cover was applied on a plane solar collector for heating water, obtaining an increase of 15 °C over a similar plane solar collector but with commercial black paint. Combining the smoke black and the elliptic geometry of tubes on a plane solar collector, its thermal efficiency increased in about 13 % as compared with a similar solar collector but without these innovations.

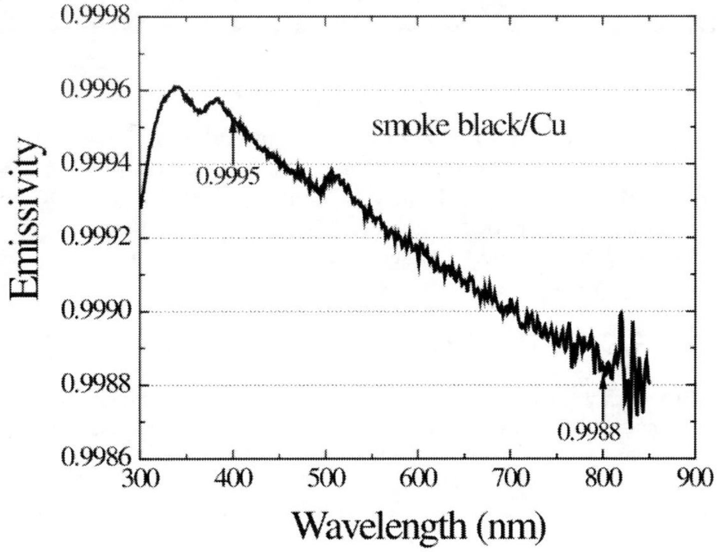

Figure 3. Emissivity (absorbance) of copper surface covered with smoke black. High emissivity (absorbance) values between 99.88% and 99.95% were measured.

Acknowledgments
Authors recognized the technical help of Victor Rejón, Emilio Corona, Radamés Espósitos, and Ulises Arias. This work was financially supported by Conacyt (México) through Innovapyme project No. 155578.

References

[1] Maldonado R D, Oliva A I, Díaz E A and Montalvo A I 2011 Analytical thermal model of a solar water heater for social housing application *Proceedings of the Second International Congress on Instrumentation and Applied Sciences* Puebla Puebla México 5-8 October
[2] Peters D W, Davids P, Wendt J R, Cruz-Cabrera A A and Kemme S A 2010 Metamaterial-inspired high-absorption surfaces for thermal infrared applications *Proceedings of the SPIE 7609 76091C Photonic and Phononic Crystal Materials and Devices X.* 26 January San Francisco California USA
[3] Anomet Inc MIRO Therm Reflect Success *http://www.anomet.com*
[4] Alanod-Solar Inc *http:/www.alanod-solar.us*
[5] Wu Ch and Shvets G 2012 *Optics Letters* **37** 308
[6] Luyo-Alvarado S and Mendez-Lango E 2008 Sistema de medición de emisividad de cuerpos negros y materiales vía reflectancia modulada *Proceedings of the Simposio de Metrología* Santiago de Querétaro Qro México 22-24 de Octubre SM2008-M221-1111
[7] Yang Z P, Ci L, Bur J, Lin S Y and Ajayam P M 2008 *NanoLetters* **8** 446

IOP Conf. Series: Materials Science and Engineering **45** (2013) 012020 doi:10.1088/1757-899X/45/1/012020

Study of extraction of Zn (II) from ammoniacal media with ionic liquids

L Hernández[1], L E Hernández[1], F Legorreta[1] and M Avila-Rodriguez[2]
[1] Área Académica de Ciencias de la Tierra y Materiales, Universidad Autónoma del Estado de Hidalgo, Carr. Pachuca-Tulancingo Km. 4.5, C. P. 42074, Mineral de la Reforma, Hgo., México.
[2] Departamento de Química Sede Pueblito de Rocha, Universidad de Guanajuato, Cerro de la Venada S/N, Guanajuato, México.

E-mail: he091802@uaeh.edu.mx

Abstract. In this paper a systematic study on liquid-liquid extraction of Zn (II) from ammoniacal media using ionic liquids (IL) as extractants is presented. The ionic liquids used for this study were: trihexyl(tetradecyl)-phophonium chloride (Cy IL 101), bis trihexyl(tetradecyl)-phophonium-(trifluoromethylsulfonyl) imide (Cy IL 109) and 1-octyl bromide pyridinium (1-BOP). With bis trihexyl(tetradecyl)- phophonium-(trifluoromethylsulfonyl) imide (Cy IL 109), the efficiency of the liquid-liquid extraction system was evaluated as a function of the nature of the aqueous phase (HCl, H_2SO_4 and NH_4/NH_3), the pH, the diluents nature (kerosene, tributyl phosphate and toluene), and the modifier content. The results showed a very fast kinetic of the Zn(II) transfer to the organic phase, reaching the equilibrium within 10 minutes of contact between the phases (organic/aqueous). On the other hand, the Zn(II) extraction was better (percentage of extraction 98% at pH 9.2), when it was extracted from NH^+_4/NH^-_3 media with an organic phase composed by Cy IL 109 dissolved in kerosene with 50% of decanol. The results showed that the uses of ILs in the extraction of metals ions, instead of organic solvents, in the liquid-liquid extraction processes are ecologically viable alternatives that require further study for its application.

1. Introduction

Several technologies can be used to remove toxic metals from liquid effluents, including precipitation, solvent extraction, ion exchange, etc. [1]. Among these technologies, solvent extraction is highly used in the recovery and separation of metals from aqueous solutions; [2]. One disadvantage of solvent extraction is the losing of organic diluents by volatilization, which has a detrimental impact on the environment and human health. Consequently, greener extraction methods are being searched; the use of ionic liquids (ILs) could overcome this disadvantage [3]. Recently, considerable interest has been manifested in the use of room temperature ionic liquids as solvents for industrial catalytic reactions [4]. This approach appears to allow the controlled production of wanted products from reactants with a minimum waste of production through side reactions due to the tendency of ionic liquids to suppress conventional solvation and solvolysis phenomena. ILs are usually composed of heterocyclic organic cations and various anions that have unique properties such as non-volatility, non-flammability, and a wide temperature range to remain in liquid phase [5]. The most attractive property of ILs according to separation techniques is that ionic liquids can be designed to be water-immiscible salts by extending the chain length of the cation and/or incorporating hydrophobic anions. These hydrophopic ILs, are available for separation techniques that use a biphasic system, like liquid-liquid extraction [6].

The purpose of this researching work is to show the feasibility of ILs as alternative to traditional organic solvents employed in liquid-liquid processes and to study the extraction of Zn(II) from sulfate media using as extractant the ionic liquids (IL), trihexyl (tetradecyl)-phophonium chloride (Cy IL 101), bis trihexyl(tetradecyl) - phophonium - (trifluoromethylsulfonyl) imide (Cy IL 109) and 1-octyl bromide pyridinium (1-BOP).

1. Experiment

1.1. Reagents

As extractant phase (organic) we used three different ionic liquids: two commercial ionic liquids, Cy IL 109, Cy IL 101, both provided by Cytec Industries, and 1-BOP (see Figure 1), which was previously synthesized by our working group. They were prepared at 0.1 mol L^{-1}, diluted in kerosene (98%, Aldrich) and n-decanol (98%, FAFC). In this case, the experimental conditions were: Organic phase=Cy IL 109, Cy IL 101, 1-BOP diluted in decanol/kerosene 50% v/v, Aqueous Phase=NH_4/NH_3 [Zn(II)]=10 ppm, pH=6.5, 7.0, 7.5, 8.0, 8.5, 9.0, 9,5, 10.0, 10.5, 11.0.

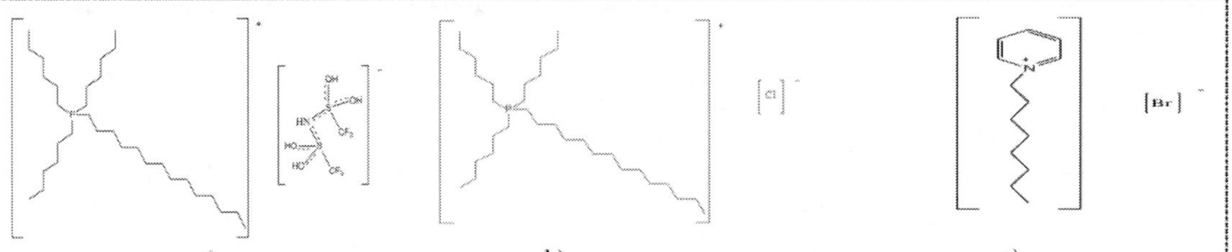

a) b) c)

Figure 1. Chemical structure of ionic liquids used. a) Cy IL 109, b) Cy IL 101, (Cytec, 2004). c) 1-BOP.

The effect of using different diluents (kerosene, tributyl phosphate (TBP (98%, Aldrich)) and toluene (98%, Reasol)) in decanol proportion: modifier, 50:50 was studied. These tests were done only with the ionic liquid Cy IL 109.

The aqueous phase was prepared from zinc sulfate ($ZnSO_4 \cdot 7H_2O$, 98%, Merck) at 10 ppm. It was diluted in deionized water in a different media: H_2SO_4 (98.08%, UT Baker), HCl (98%, Aldrich), NH_3 (analytical grade 25% v / v, Aldrich) and NH_4Cl (analytical grade, J. T. Baker).

Kinetic study was made only with Cy IL 109 because it was the IL that provided the best results. This one was diluted in decanol / kerosene 50% v / v, using like aqueous phase a 0.01mol L^{-1} ammoniacal solution (pH 9.2).

1.2. Methodology

The Zn(II) extraction studies with different ionic liquids and diluents were carried out by contacting equal volumes of both phases (aqueous and organic) in an erlenmeyer flask, which was placed in a ping pong shaker (Eberbach model 6000), by 120 minutes. Subsequently the two phases were separated (in a separatory funnel) and the content of Zn(II) into the aqueous phase was analysed by atomic absorption spectrophotometry (AA) with a Perkin-Elmer 3110 (EAA) at a 213.9 nm wavelength. The Zn(II) concentration in the organic phase was obtained by mass balance.

In the kinetic study, it was used a double layer glass reactor, placing equal volumes of both phases (aqueous and organic). The mix was kept stirred in a magnetic stirring grid at constant temperature (298 K). After each test, the phases were separated for their subsequent Zn(II) analysis by AA. In the experiments, the pH of the aqueous solutions was measured using an automatic titrator (716 DMS Titrino Metrohm) with a glass electrode.

2. Results and Discussion

2.1. Effect of different ionic liquids

The results obtained in the Zn(II) extraction with three different ionic liquids, Cy IL 109, Cy IL 101 and 1-BOP, as a function of pH are showed in the figure 2. It is possible to see that the pH drastically influences on the extraction of the metal ion. There is a high zinc extraction of 98% at pH 9.2 with Cy IL 109 and 69% at pH = 3.2 with Cy IL 101. The Cy IL 101 extracts better in an acid medium, otherwise occurs with Cy IL 109 that has a better performance in basic media. This phenomenon is attributed to the nature of the different anions containing in each ionic liquid. Regarding to the ionic liquid 1-BOP, we could obtain 100% extraction at pH 7.6. However, this compound showed high solubility in water and for this reason it does not accomplish the required characteristics to be considered as a good solvent or IL [9].

3.2 Effect of the aqueous media

The results of the Zn(II) extraction from H_2SO_4 and HCl media showed that there is no a relevant extraction of this ion, reaching a maximum extraction percentage of 9.5% to pH 2.0 when H_2SO_4 was used, and 10.8% with HCl at pH 2.4(see Fig. 3). The behavior of the system did not show a stable trend, which can be attributed to the appearance of a third phase. Nevertheless, it is possible to say that the extraction of Zn(II) by the Cy IL 109 form acidic media is inefficient.

When we worked the aqueous phase in an ammoniacal solution the Zn(II) extraction was better than the one obtained in an acidic media.

It is necessary to remark that a second organic phase is formed during the liquid-liquid extraction of Zn(II), which is attributed to the insolubility of Cy IL 109 in kerosene. The presence of n-decanol, besides helping to have a good phase separation, it helps to solubilise the Cy IL 109 in the organic phase. However, in relation kerosene / decanol (90:10) the solubility is still limited. When the organic phase is only composed by kerosene, the insolubility of ionic liquid is evident. In fact, we verified the presence of Zn (II) in the second organic phase, which agrees with Dingsheng et al [7].

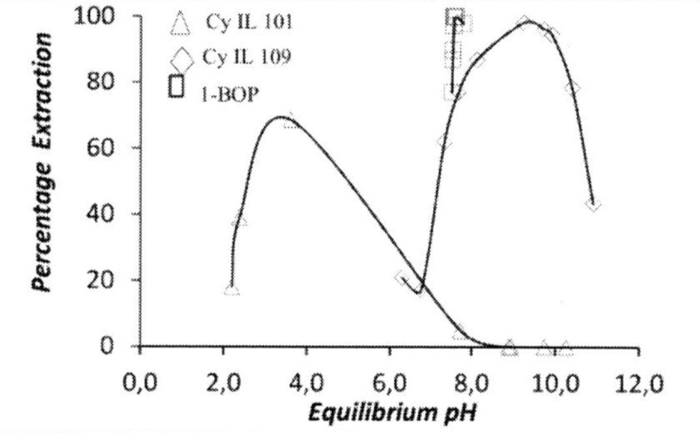

Figure 2. Effect of equilibrium pH on the extraction of Zn (II) using different ionic liquids.

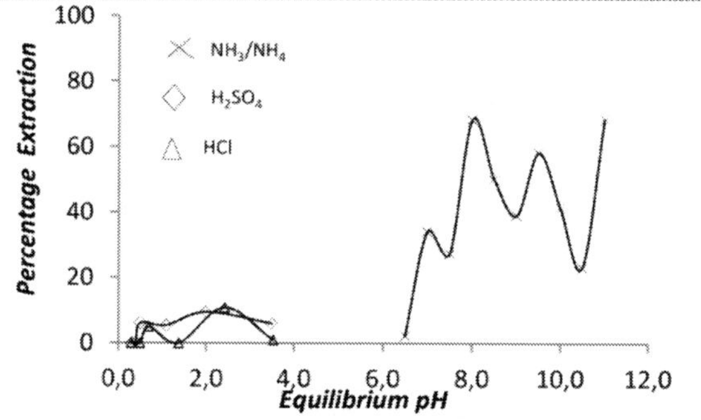

Figure 3. Effect of equilibrium pH on the extraction of Zn (II) in different media.

3.3 Effect of diluent

In solvent extraction is often added to the organic phase a third component called "modifier" [8], in order to prevent the formation of a third phase or the presence of stable emulsions in the extraction systems. In this sense, we studied the behavior modifier dissolved in three different diluents (see Fig. 4), kerosene, tributyl phosphate and toluene in decanol (proportion: 1:1). The results showed that with TBP and toluene was possible to get extractions of 99 and 97% at pH 10.4 and 8.4, respectively. When kerosene is used as diluent, the Zn(II) extraction was of 98% at pH 9.2, comparing it with that reported by Regel-Rosocka [10]. The Cy IL 109 in toluene extracted only 20% of Zn(II) from HCl solutions.

3.4 Effect of decanol/kerosene proportion

The Zn(II) extraction from ammoniacal media at pH 9.2, using Cy IL 109 as extractant, at different kerosene / decanol rate, shows that the Zn(II) extraction percentage increases as the relation of decanol / kerosene increases, reaching percentage of extraction up to 98% using a rate decanol / kerosene 1:1, (see Fig. 5).

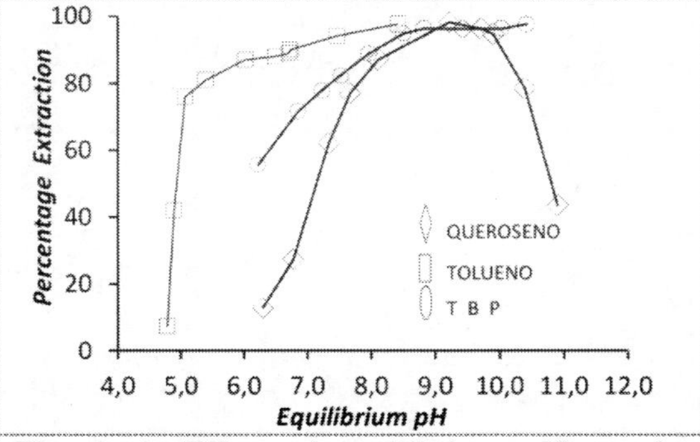

Figure 4. Effect of equilibrium pH on the extraction of Zn (II) using Cy IL109 and three different diluents

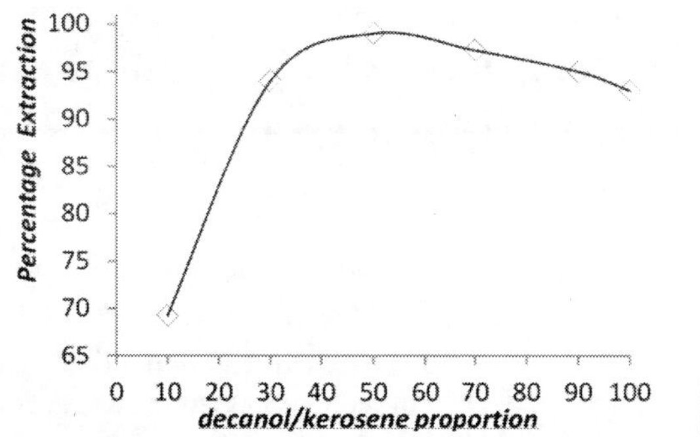

Figure 5. Effect of Zn (II) extraction at different decanol/kerosene proportion.

3.5 Effect of time

The results obtained from the kinetic study, in the case of the Zn(II) by the Cy IL 109, showed that after 10 minutes of stirring it was possible to obtain 81% of Zn (II), this value increased until 160 minutes, time when we achieved a maximum extraction percentage (98%) which was maintained constant.

3. Conclusions

The liquid-liquid extraction of Zn (II) using Cy IL 109 as extractant dissolved in kerosene and n-decanol was possible in an ammoniacal medium (98% at pH 9.2). When the Cy IL 101 was used the Zn(II) extraction from acidic media was possible. When the 1-BOP was used as extractant the extraction of Zn(II) was very high, but the use of this extractant it is not recommended because of its high solubility in water. The most suitable conditions for the extraction of Zn (II) are: Cy IL 109 diluted in decanol-kerosene ratio of 1:1 at pH 9.2. The use of ionic liquids could be competitive and in the future they could be applied to systems containing different metal species.

4. References

[1] Lozano L J and Godínez C 2003 Comparative Study of SX of Vanadium from Sulphate Solutions by Primene 81R and Alamine *Miner Eng* **16** 291-294

[2] Agrawal A Kumari S and Sahu K K 2008 Liquid-Liquid Extraction of H_2SO Afrom Zinc Bleed Stream *Hydrometallurgy* 42–47

[3] Rogers R D and Voth G A 2007 Ionic Liquids *Acc. Chem. Res* **40** 1077–1078

[4] Wilkes J S 2004 Properties of IL Solvents for Catalysis *J. Mol. Catal. A Chem* 11-17

[5] Kazunori N 2005 *Ind. Eng. Chem. Res* **44** 4368-4372

[6] Luo H., Dai, S. and Bonnesen P.V 2004 *Anal. Chem* **76** 2773-2779

[7] He X 2006 Study of transport and separation of Zn(II) by a combined SLM strip dispersion process containing D2EPHA in kerosene as the carrier *Dasalination* 40-51

[8] Juaír D Meseguer D 1998 Extracción de cromo con disolventes orgánicos. Extracción con DEHPA *Rev. Metal* **3** 253-260

[9] Papaiconomou N Lee J L 2007 Selective Extraction of Copper, Mercury, Silver and Palladium Ions from Water Using Hydrophobic Ionic Liquids *American Chemical Society* A-G

[10] Regel-Rosocka M 2008 Extractive removal of Zinc(II) from chloride liquors with phosphonium ionic liquids/toluene mixtures as novel extractants *Separation and Purification Technology* 960-965

IOP Conf. Series: Materials Science and Engineering **45** (2013) 012021 doi:10.1088/1757-899X/45/1/012021

Evaluation of dyes adsorption properties of TiO_2-alginate biohybrid material.

J A Barrón Zambrano[1], G Sánchez Morales[2], A Ávila Ortega[1], D Muñoz Rodríguez[1] y C Carrera Figueiras[1]

[1]Universidad Autónoma de Yucatán, CCEI, Facultad de Ingeniería Química.
[2]Universidad Veracruzana, Facultad de Ciencias Químicas

E-mail: jesus.barron@uady.mx

Abstract. In this study a TiO_2-alginate biohybrid material was obtained by the sol gel method and its adsorption properties were compared to those of its precursors using eosin B (anionic) as model dye. The results showed that the TiO_2 and biohybrid have a greater affinity for eosine B than alginate. The maximum adsorption capacity for the eosin B was obtained at pH = 10. Kinetic studies showed that the biohybrid has greater rate and adsorption capacity than its precursors. Kinetic data were fitted to a pseudo-second order kinetic model. The experimental isotherms were fitted to the Langmuir model.

1. Introduction

Effluents contaminated with dyes generated from various industrial operations are important because they contain several toxic compounds that affect aquatic life because of their high biological and chemical oxygen demand. Today the removal of this kind of pollutants requires more precise and effective techniques, such as adsorption, which has a high capacity for the purification of effluents [1].

In recent years there has been a growing interest in the development of new hybrid materials and their potential application as adsorbents for the removal of different types of pollutants. In this sense, polymethacrylic acid (PMAA) was grafted onto the surface of silica gel particles (PMAA/SiO_2) for the adsorption of phenol [2].

Some authors have investigated the feasibility of using TiO_2 based materials as new adsorbents for solid phase extraction (SPE) of organic compounds, like DDT and its metabolites, and metals in water [3-6]. TiO_2 has different selectivity for a variety of analytes when compared to a silica matrix. Therefore, the use of TiO_2 as inorganic phase instead of silica, could serve to improve the selectivity of the adsorption processes [4]. In other work, TiO_2 particles coated with polyacrylonitrile, was used for the extraction of metals, with recovery rates from 60 to 95% [7].

On the other hand, chitosan-silica hybrid materials obtained as microparticles had been prepared by the sol-gel method using TEOS or oligomers in the presence of the biopolymer. These materials may be used as stationary phase in HPLC [8]. The hybrid particles, based on algal polysaccharides such as alginate and carrageenan, appear to be a good candidate for the formation of silica-based biocomposites. Alginate is a copolymer of L-guluronic and D-mannuronic acid that can be reversibly gelled by the addition of divalent metal, like calcium [9]. Chuang et al. [10] used bamboo and modified bamboo on its surface with TiO_2 nanoparticles for the removal of benzene and toluene, obtaining good results.

In these examples, the interaction of the biological and inorganic components have synergistic effects that lead to hybrid materials with improved mechanical strength, higher thermal and chemical stability, and in some cases, functional properties.

In this study TiO_2-alginate biohybrid material was obtained using the sol gel method and the adsorption properties were compared to their predecessors using an anionic dye model: eosin B (EB). The effect of pH, contact time and initial concentration of dye on the adsorption capacity of different materials were studied. The kinetic parameters were also calculated to determine rate constants. The equilibrium experimental data were fitted to Langmuir model.

2. Materials and Methods

2.1. Synthesis of adsorbents

TiO_2 was synthesized by mixing Titanium Isoproxide (TIPO), ethanol and distilled water in a beaker with slow stirring. 1 M HCl was added to the mixture until pH 2 and allowed to stir for 30 minutes. 0.1 M NaOH was added until pH 7 and then continued with stirring to obtain a gel. The gel obtained was allowed to grow for 24 h, and then washed with water. Subsequently, the product was dried at 85 °C in air oven to constant weight.

Calcium alginate powders were obtained by spraying 600 mL of 2% sodium alginate solution on 500 mL of 0.05 M $CaCl_2$ solution. The formed calcium alginate particles were allowed to mature for 24 h. After that, they were filtered, washed with distilled water and dried at 50 °C in air oven for 24 h. After drying, the particles were maintained in a desiccator.

Biohybrid sorbent was synthesized by mixing TIPO, ethanol and distilled water in a beaker with slow stirring. The ratio TIPO/H_2O was kept constant and equal to 1. Then 1 M HCl was added to the mixture until pH 2 and allowed to stir for 30 minutes. The above mixture was added dropwise to 500 mL of a solution of 2% sodium alginate with rapid stirring until total homogenization of the two solutions. After completed addition, agitation was reduced and continued to obtain a gel. Subsequently, the gel was dried in vacuum oven at 85 °C and washed with distilled water, then dried again at 85 °C in air oven to constant weight. All materials were characterized by BET technique using a Quantachrome ASiQwin analyzer.

2.2. Effect of initial pH

30 mL of dye solution of 50 mg/L initial concentration different pH values (1 to 11) were put in contact with 0.1 g of sorbent and agitated for 24 h. The resultant solutions were filtered before analyses. Initial and equilibrium dye concentrations were determined by spectrophotometry using UV-VIS spectrophotometer (Thermo Scientific, LR 162800). Absorbance measurements were made at the maximum wavelength of EB (λ= 517 nm). The amount of adsorbed EB, q_e (mg/g), was calculated from mass balance using equation (1):

$$q_e = \frac{(C_0 - C_e)V}{m} \quad (1)$$

where C_0 and C_e are the initial and equilibrium concentrations of dye solution (mg/L), respectively, V is the volume of dye solution (L) and m is the mass (g) of the sorbent used.

2.3. Sorption kinetics

150 mL of EB solution (C_o= 50 mg/L) at the pH of maximum adsorption, which was previously determined, was put in contact with 0.5 g of sorbent and placed under stirring for 10 h. Samples were taken at predetermined time intervals and analyzed as described before. Kinetic data were analyzed with a pseudo-second-order model (equation (2)) [11].

$$\frac{t}{q_t} = \frac{1}{K_2 q_e^2} + \frac{1}{q_e} t \quad (2)$$

where q_e is the adsorption capacity at equilibrium (mg g^{-1}), q_t is the adsorbed amount of sorbate at time t (mg g^{-1}), and K_2 is the pseudo-second order rate constant (g mg^{-1}min^{-1}).

2.4. *Adsorption isotherms*

0.1 g of sorbent was added to a bottle containing 30 mL of EB aqueous solution of desired concentration (10-100 mg/L) at the pH of maximum adsorption, previously determined. The bottles were subsequently capped and placed on a mechanical shaker at a speed of 200 rpm at (20±2) °C for 24 h. Initial and equilibrium dye concentrations were determinate as describe above. Experimental data were fitted to the isotherm model of Langmuir (equation (3)) [10].

$$q_e = \frac{q_{max} K_L C_e}{1+K_L C_e} \tag{3}$$

where C_e is the supernatant concentration at the equilibrium (mg L^{-1}), K_L is the Langmuir affinity constant (L mg^{-1}), and q_{max} is the maximum adsorption capacity of the sorbent (mg g^{-1}) assuming a monolayer coverage of the sorbent.

3. Results and discussions

3.1. *pH Influence*

Figure 1 shows the influence of pH on the sorption of eosin B onto on TiO_2, TiO_2/Alginate biohybrid, and alginate. It can be see that both TiO_2 materials have great affinity for EB. As expected, alginate has a very low sorption capacity, due to his negative surface charge, that repels the anionic EB. This results shows that the biohybrid keeps the adsorption properties of TiO_2.

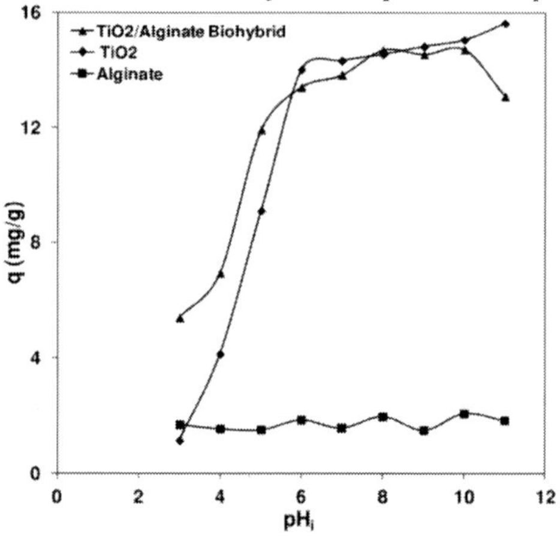

Figure 1. Influence of pH on the sorption of eosin B on TiO_2, TiO_2/Alginate biohybrid, and alginate.

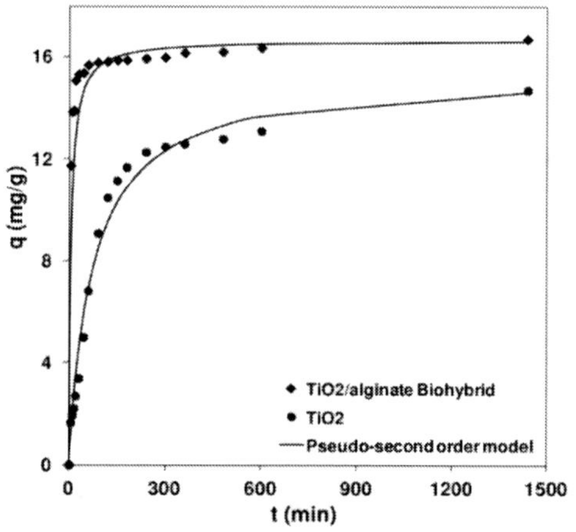

Figure 2. Sorption kinetics of eosin B on TiO_2/Alginate biohybrid and TiO_2.

3.2. *Sorption kinetics*

Figure 2 shows that the rate of sorption of the biohybrid is greater than that of the TiO_2. That seems to indicate that the alginate increases the porosity of the adsorbent, making the active sites of the biohybrid more accessible for the molecules of EB. BET analysis shows that the biohybrid has a half pore width of 16.6 A, while for the TiO_2 it is 8 A. Experimental data fitted well to the pseudo-second order model ($R^2 > 0.99$). The values of K_2 for the biohybrid and TiO_2 were 0.009 and 0.00085 respectively, which agree with the experimental observations.

3.3. *Sorption isotherms*

Sorption isotherms for the biohybrid and TiO_2 are depicted in figure 3. It can be see that the both experimental isotherms fit well with the Langmuir model. The values of q_{max} were 263.7 mg/g for the biohybrid and 398.3 mg/g for the TiO_2. Although the TiO_2 seems to have a higher adsorption capacity that the biohybrid, it must be pointed out that the biohydrid has only half the mass of TiO_2, and that the alginate contributes little to the total capacity of the biohybrid, so his adsorption capacity is higher than that of the TiO_2.

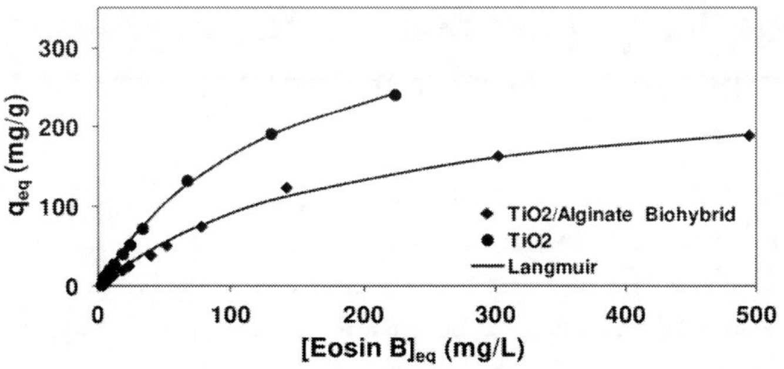

Figure 3. Sorption isotherms of eosin B on TiO_2/Alginate biohybrid and TiO_2

4. Conclusions

A new TiO_2/alginate biohybrid was synthesized by sol-gel method and his dye adsorption properties evaluated and compared to that of his precursors using an anionic eosin B, as model dye. The results show that pH has a great influence on the dye sorption onto the different studied adsorbents. Maximum dye adsorption capacity for the biohybrid was obtained a pH = 10. Kinetic and equilibrium data shows that the biohybrid has higher rate and adsorption capacity than his precursors. Kinetic data were well described by the pseudo-second order kinetic model, while equilibrium data fitted well to the Langmuir model. The improved rate and adsorption capacity of biohybrid shows a synergistic effect of the two components; from titanium oxide the biohybrid acquired the property of adsorbing anionic molecules, while the alginate yielded a more porous structure.

5. Acknowledgements

The authors thank the Consejo Nacional de Ciencia y Tecnología (CONACyT) for financial support to carry out this work through the basic science project 83390

References

[1] Cestari AR, Vieira EFS, Vieira GS,. da Costa LP, Tavares AMG, Loh W and Airoldi C 2009 *J. Hazard. Mater.* **161** 307
[2] An F, Gao B and Feng X 2009 *Chem. Eng. J.* **153** 108
[3] Zhou Q-X, Ding Y, Xiao JP, Liu G and Guo X 2007 J. Chromatogr. A **1147** 10; Zhou Q-X, Zhao X-N and Xiao JP 2009 *Talanta* **77** 1774
[4] de Moraes SVM, Passos JB, Schossler P, Caramao EB, Moro CC, Costa TMH and Benvenutti EV 2003 *Talanta* **59** 1039
[5] Engholm-Keller K and Larsen MR 2011 *J. Proteomics* **75** 317
[6] Ma L and Tu SX 2010 *Environ. Chem. Lett.* **9** 465
[7] Ali IM, El-Zahhar AA and Zakaria ES 2005 *J. Radioanal. Nucl. Chem.* **264** 637
[8] Rashidova S, Shakarova D, Ruzimuradov O, Satubaldieva D, Zalyalieva S, Shpigun O 2004 *J. Chromatogr.* **800** 49
[9] Xu B, Bartley J P and Johnson, R A 2003 *Appl. Polym. Sci.* **90** 747; Veglio F, Esposito A and Reverberi A 2002 *Hydrometallurgy* **65** 43
[10] Chuang CS, Wang M-K, Ko C-H, Ou C-C and Wu C-H 2008 *Bioresour. Technol.* **99** 954
[11] Ho YS, McKay G 1998 *Chem. Eng. J.* **70**, 115-124

Matrix solid-phase dispersion extraction of organophosphorus pesticide using SiO_2-poly(N-vinylimidazole)

M C Gutiérrez-Solís[1], D Muñoz-Rodríguez[1], S Medina-Peralta[2], C Carrera-Figueiras[1], A Ávila-Ortega[1]

[1] Facultad de Ingeniería Química, Campus de Ingenierías y Ciencias Exactas, Universidad Autónoma de Yucatán, Mérida, Yucatán, México.

[2] Facultad de Matemáticas, Campus de Ingenierías y Ciencias Exactas, Universidad Autónoma de Yucatán, Mérida, Yucatán, México.

E-mail: david.mr@uady.mx

Abstract. A sorbent material based on silica particles modified with poly(N-vinylimidazole) (SiO_2-PVI) has been evaluated for the treatment of samples by matrix solid-phase dispersion (MSPD). The extraction of four organophosphorus pesticides was done from a spiked tomato and the extracts were analyzed by gas chromatography coupled to mass spectrometry. Six elution solvents were evaluated and acetone was selected due to better recovery of the four pesticides and low background signal in the chromatograms. A factorial design 2^4 was used for selection of extraction conditions. The factors were contact time, acetone volume, treatment (with or without freeze-drying) and adsorbent (SiO_2 or SiO_2-PVI). The best recoveries were obtained using 15 minutes of contact, 2 mL of solvent and sorbent without freeze-drying. The recoveries were between 60 and 83% for SiO_2-PVI in spiked tomato with 0.2 and 0.8μg/g.

1. Introduction

The development, evaluation and application of new adsorbents in matrix solid-phase dispersion (MSPD) are scarce. The current adsorbents based on organic functionalized silica have a relatively low retention of polar compounds and low chemical stability in extreme pH conditions ($pH<2$ and $pH>8$)[1, 2]. The retention can improve by modification the surface of sorbents with polar groups, for example, the poly(N-vinylimidazole) to get polar interactions[3,4,5]. In recent years there has been a growing interest in the development of new hybrid materials and their application as adsorbents for the removal of different types of pollutants. In this sense, polymethacrylic acid (PMAA) was grafted onto the surface of silica gel particles (PMAA/SiO_2) for the adsorption of phenol [6].

Researchers have reported studies on compounds derived from silica, which have been successfully applied as adsorbents in the removal of pollutants such as aromatic compounds and some heavy metals. An et al. [7] researched macromolecules grafted to polyacrylamide (PAM) on the surface of particles of silica gel using 3-methacryloxypropyl trimethoxysilane (MPS) as coupling agent. They developed a new adsorbent material PAM/SiO_2 for removal of 2,4,6-trinitrotoluene (TNT).There have also been studies on their use as columns for the detection of some heavy metals using the method of solid phase extraction.

Organophosphorus pesticides (OPPs) are synthetic substances most used to protect crops and hives. Nevertheless, OPPs are toxic to human being. Exposure to even small amounts of OPPs can be fatal

Published under licence by IOP Publishing Ltd

and death is usually caused by respiratory failure. For this reason, analysis of OPPs residues is necessary to protect consumer's health and guarantee the quality of vegtables like tomato [8].

In this work, a sorbent material based on silica particles modified with poly(N-vinylimidazole) shell (SiO_2-PVI) has been evaluated for the treatment of samples by matrix solid phase dispersion (MSPD). The extraction of four organophosphorus pesticides (parathion-methyl, malathion, dichlofenthion and fensulfothion) was done from a fortified tomato and the extracts were analysed by gas Chromatography coupled to mass spectrometry.

2. Experimental

2.1. Chemicals and materials

SiO_2 from Sigma-Aldrich and SiO_2-PVI was obtained as previously reported. Analytical grade anhydrous sodium sulfate (Na_2SO_4) and solvents (ethyl ether, dichloromethane, acetonitrile, ethyl acetate, acetone, 2-propanol, isooctane) were from J. T. Baker (Mallinckrodt Baker, Phillipsburg, NJ, USA). Pesticide standards (at least 97% pure) [methyl parathion (MPT), malathion (MLT), dichlofenthion (DCF) and fensulfothion(FSF)] were from ChemService (West Chester, PA, USA).

Individual stock solutions were prepared in ethyl acetate (1000 ppm) and stored at -18°C. Working standard solutions were prepared by diluting the stock solutions with ethyl acetate. Finally, calibration standards were prepared by diluting the stock solution with the extract of tomato free of pesticides.

2.2. Spiking procedure

The preparation of the spiked tomato was as follows: 200μL of a standard mixture (0.1 or 0.4 μg/mL) of organophosphorus pesticides (MPT, MLT, DCF, FSF) were added to 0.1 g of tomato respectively to reach a concentration of 0.2 and 0.8 μg/g. Finally, it was left to stand (15 min) to allow the incorporation of the pesticides in the matrix of the tomato.

2.3. Experimental design

An unreplicated factorial design 2^4 for each pesticide was used and four factors with two levels were evaluated: (1) contact time (A: 0 min, 15 min), (2) volume of solvent (B: 1 mL, 2 mL), (3) particles treatment (C: without freeze-dried, with freeze-dried), (4) type of adsorbent (D: SiO_2, SiO_2-PVI). The recovery percent (%R) of pesticides in tomato extracts was the variable response. The results were interpreted using an analysis of variance (ANOVA) [9].

2.4. MSPD extraction

First, a 0.1 g of tomato was placed in a glass mortar. Immediately, 0.5g of Na_2SO_4 was added and disgregated with the pestle. Next, 0.4 g of the particles (SiO_2 or SiO_2-PVI) were added and dispersed to obtain an homogenous mixture. Subsequently, it was transferred to a polypropylene column (100 x 20 mm ID) with a filter paper disc at the bottom. After that, the mixture was compacted slightly to remove air bags. The end of the column was blocked with a polypropylene cap and 2 mL of acetone were loaded. Then, the entrance of the column was closed and the mixture with the solvent was left to stand (15 min); it was eluted dropwise. The eluent was evaporated to 1 mL with gentle air flow and 0.1 g of Na_2SO_4 was added. Finally, the sample was centrifuged (14500 rpm, 4 min) and the supernatant was transferred into a vial for GC-MS analysis.

2.5. Instrumentation and chromatographic conditions

The extracts of the pesticides were analysed with a gas chromatograph (6890N, Agilent Technologies) coupled to a mass spectrometer (5973 N, Agilent Technologies). Helium (purity 99.999) was used as carrier gas (1 mL/min). A fused-silica column Equity-5™ (5% phenyl-95% polydimethylsiloxane; 30 m × 0.25 mm ID, 0.25 μm), supplied by SUPELCO (Bellefonte, PA, USA) was employed. The inlet temperature was 250°C and 1 μL of the extract was injected in a splitless mode (0.7 min). The oven temperature as programmed as follows: 120°C (initial temperature), increase of temperature to 280°C

IOP Conf. Series: Materials Science and Engineering **45** (2013) 012022 doi:10.1088/1757-899X/45/1/012022

at 10ºC/min and holding 4 min. The eluent from the GC column was transferred into the MSD via a transfer line (at 280ºC). Typical conditions were optimized through the autotune software option. The electron impact mode (70eV) was used as ionization source (250ºC) and masses were monitored between 50-400 m/z (SCAN mode). The temperature of the quadrupole was 150ºC. The analysis was performed in the selected ion monitoring (SIM). The mass-charge ratio (m/z) for the ions monitored of each pesticide were 97, 223, 279 for DCF; 109, 125, 263 for MPT; 125, 158, 173 for MLT and 265, 293, 308 for FSF.

2.6. Efficiency of the MSPD extraction with SiO_2-poly(N-vinylimidazole)

Recovery experiments were done at two spiked levels (0.2 and 0.8 μg/g) to evaluate the efficiency of the SiO_2-poly(N-vinylimidazole) in MSPD. Recoveries were calculated through external calibration and pesticide standards were prepared by dilution with bee pollen extracts to minimize matrix effect. The precision of extractions was evaluated as the relative standard deviation (RSD). Tomato at each spiked level was analysed at least twice.

3. Results and discussion

The chromatogram obtained from a mixture of pesticides is showed in figure 1. It can be seen that all the chromatographic signals were resolved at the base line.

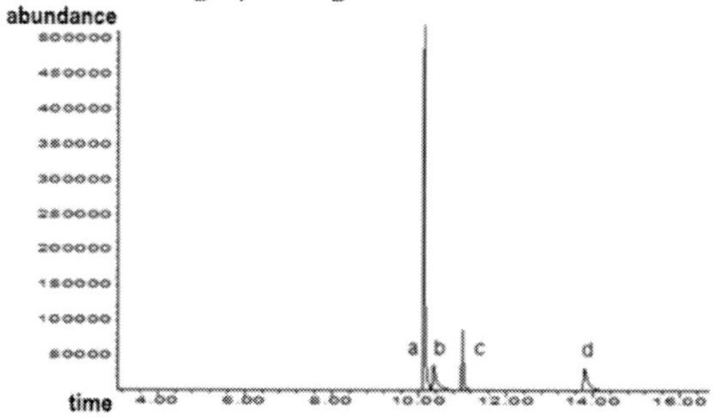

Figure 1. GC/MS-SCAN chromatogram acquired from a standard mixture of organophosphorus pesticides to 2 μg/mL. a: DCF, b: MPT, c: MLT, d: FSF

Solvent	Relative analytical signal for pesticides			
	DCF	**MPT**	**MLT**	**FSF**
Ethylacetate	0.35	0.02	0.12	0.04
Ethyl ether	0.20	0.96	0.47	0.10
Dichlometane	0.33	1.00	0.82	0.01
2-Propanol	0.37	0.14	0.29	1.00
Acetonitrile	0.59	0.03	0.17	0.56
Acetone	1.00	0.88	1.00	0.99

n=2

Table 1. Relative analytical signal for the pesticides (0.5 μg/g) obtained from the tomato extract in different solvents and analyzed by GC/MS-SIM.

Dichloromethane, ethyl ether, ethyl acetate, acetone, 2-propanol and acetonitrile were tested for the elution of pesticides. As is showed in table 1, acetone was chosen as the solvent of elution because of the high relative analytical signal achieved for pesticides. In addition, chromatograms acquired from acetone extracts presented a lower number of chromatographic signals as is showed in figure 2.

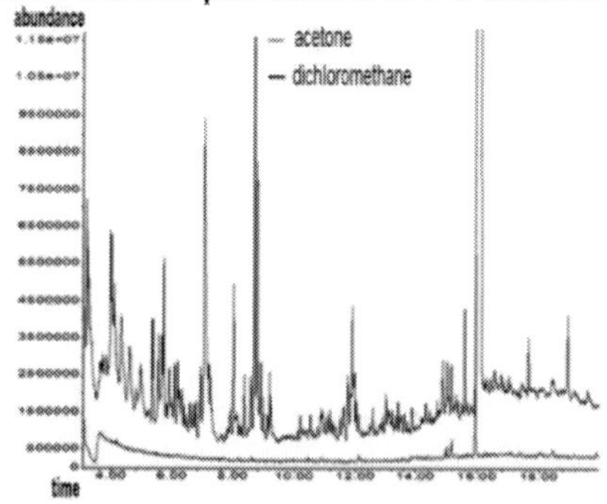

Figure 2. GC/MS-SCAN chromatogram acquired from extracts of tomato with two different solvents.

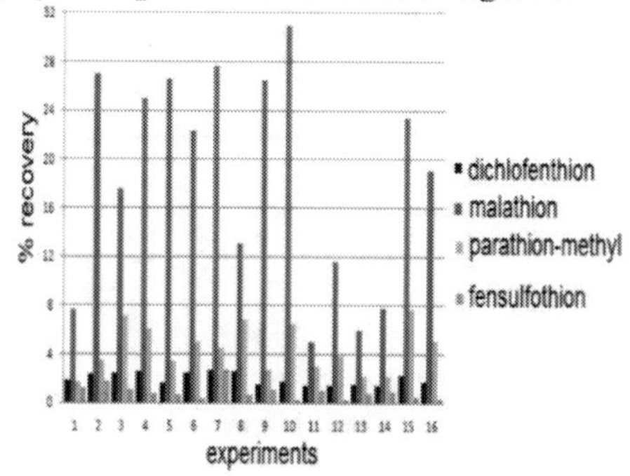

Figure 3. Recoveries obtained in the 16 experiments for each pesticide, using SiO_2-PVI as adsorbent.

Once the solvent elution was selected, an experimental design was used to find the combination of extraction conditions to extract pesticides from spiked tomato by MSPD with materials based on silicon. The graph of the recoveries obtained for three pesticides in each experiment is observed in figure 3. The best extraction conditions were: 15 min of contact time, 2 mL of solvent volume, adsorbent SiO_2-PVI without lyophilize (experiment 15). These conditions gave to the best recovery for DCF and MPT and the fourth better recovery for FSF.

The calibration process was done with matrix-match standard solutions at 0.0, 0.01, 0.02, 0.04, 0.06, 0.08, 0.10 and 0.12 μg/mL. In table 2, the calibration function and the coefficient of determination (R^2) are showed. In all cases, the determination coefficient was higher than 0.99, which indicates a good linearity.

Table 2. Recoveries and relative standard deviations (RSD) for the pesticide extraction by MSPD using SiO_2-PVI

Pesticide	Calibration function	R^2	Spiked level, μg/g	Recovery, %	RSD, %
DCF	A = 1000000C - 3902.7	0.9985	0.2	68	12
			0.8	83	13
MPT	A = 266753C - 834.47	0.9977	0.2	73	11
			0.8	77	9
MLT	A = 665493 - 5061.3	0.9961	0.2	60	18
			0.8	68	11
FST	A = 403883C - 180.71	0.9974	0.2	73	13
			0.8	78	9

A: analytical signal (peak area); C: pesticide concentration (μg/mL)

Recoveries obtained for the organophosphorous pesticides by MSPD extraction with SiO_2-poly(N-vinylimidazole) were at least 60%. The highest recovery was achieved with DCF at 0.8 μg/g followed by MPT and FST that presented recoveries higher than 70% at both spiked level. The lowest recoveries were obtained for MLT at both spike levels and for DCF at 0.2 μg/g. For most of the pesticides the presicion of the extraction was lower than 20%.

Recoveries achieved in our work agrees with results of Torres and co-workers [10] for MLT, MPT and diazinon in tomato (spike level: 0.1 μg/g) by MSPD with C18 and ethyl acetate as solvent elution. However, our values are relatively low in comparison with those obtained by Menezes and co-workers [11] for MPT (95%) from tomato (spike level: 1.0 μg/g) with alumina and dichloromethane.

4. Conclusions.

A material based on SiO_2-PVI was applied for the extraction of organophosphorous pesticides from spiked tomato by MSPD. Recoveries and precision were acceptable (60%, RSD<20%) at two spiked levels this suggest that SiO_2-PVI has potential to be applied as adsorbent in MSPD.

References

[1] García M, Canosa P and Rodríguez I 2008 *Anal. Bioanal. Chem.* **391** 963.
[2] Gilar M, Bouvier E and Compton B 2001 *J. Chromatogr. A.* **909** 111.
[3] Fontanals N, Galià M, Marcé R and Borrull F 2004 *Chromatographia* **60** 511.
[4] Fontanals N, Galià M, Marcé R and Borrull F 2004 *J. Chromatogr. A.* **1030** 63.
[5] Fontanals N, Marcé R and Borrull F 2007 *J. Chromatogr. A.* **1152** 14.
[6] An F, Gao B and Feng X 2009 *Chem. Eng. J.* **153** 108.
[7] An F, Feng X and Gao B 2009 *J. Hazard. Mater.* **168** 352; An F, Feng X, and Gao B 2009 *Chem. Eng. J.* **151** 183.
[8] Gloria A, Salvador M, Yolanda M, David M 2011 *Anal Bioanal Chem.***400** 885.
[9] Montgomery DC 2004 *Diseño y análisis de experimentos*. 2nd Ed.(México, D.F: Limusa Wiley)
[10] Torres C, Picó Y, Redondo M and Mañes J 1995 *Chromatographia* **41** 95.
[11] Menezes A, Navickiene S and Dorea H 2006 *J. Braz. Chem. Soc.* **17** 874.

Carbonaceous material production from vegetable residue and their use in the removal of textile dyes present in wastewater

A A Peláez-Cid[1]*, M A Tlalpa-Galán[1], and A M Herrera-González[2]
[1]Facultad de Ingeniería, BUAP, Edificio 108A, Ciudad Universitaria, CP. 72570, Puebla, Puebla, México. Phone: +00 222 2295500 ext. 7632.
[2]Centro de Investigaciones en Materiales y Metalurgia, UAEH, Hidalgo, México.
E-mail: alalpeci@hotmail.com

Abstract. This paper presents the adsorption results of acid, basic, direct, vat, and reactive-type dyes on carbonaceous adsorbent materials prepared starting off vegetable residue such as *Opuntia ficus indica* and *Casimiroa edulis* fruit wastes. The adsorbents prepared from *Opuntia ficus indica* waste were designated: TunaAsh, CarTunaT, and CarTunaQ. The materials obtained from *Casimiroa edulis* waste were named: CenZAP, CarZAPT, and CarZAPQ. TunaAsh and CenZAP consist of ashes obtained at 550 °C; CarTunaT and CarZAPT consist of the materials carbonized at 400 °C; lastly, CarTunaQ and CarZAPQ consist of chemically activated carbons using H_3PO_4 at 400 °C. Only the chemically activated materials were washed with distilled water until a neutral pH was obtained after their carbonization. All materials were ground and sieved to obtain a particle size ranging from 0.25 to 0.84 mm. The static adsorption results showed that both ashes and chemically activated carbon are more efficient at dye removal for both vegetable residues. For TunaAsh and CarTunaQ, removal rates of up to 100% in the cases of basic, acid, and direct dyes were achieved. Regarding wastewater containing reactive dyes, the efficiency ranged from 60 to 100%. For vat effluents, it ranged from 42 to 52%. In the case of CenZAP and CarZAPQ, it was possible to treat reactive effluents with rates ranging between 63 and 91%. Regarding vat effluents, it ranged from 57 to 68%. The process of characterization for all materials was done using scanning electron microscopy and infrared spectroscopy.

Keywords. *Opuntia ficus indica* fruit waste, *Casimiroa edulis* fruit waste, Carbonaceous adsorbents, Textile dyes removal.

1. Introduction

Activated carbon (AC) is one of the most commonly used adsorbents in the removal process of industrial pollutants, organic compounds, heavy metals, herbicides, and dyes, among many others toxic compounds. There are two types of activation procedures to produce AC: Physical or thermal, and chemical. In physical activation, the lignocellulosic material as such or the previously carbonized materials can undergo gasification with water vapor, carbon dioxide, or the combustion gases produced. Chemical activation consists of the impregnation with chemicals such as $ZnCl_2$ or H_3PO_4 on lignocellulosic or carbonaceous raw materials, later they undergo carbonization or pyrolysis, and finally, they are washed to eliminate the activating agent. The application of a gaseous stream such as air or nitrogen is a common practice during pyrolysis which generates a better development on the porosity of the material [1].

Commercially, AC is prepared by means of oxidative pyrolysis starting off soft and hardwoods, peat, lignite, mineral carbon, bones, and coconut shell [2]. AC is considered an expensive material because of the chemical and physical treatments used during its production, its low yield, its production's high energy consumption, and the thermal or solvents treatments used for its regeneration. Attempting to reduce the cost of production of AC, contemporary research is taking a turn towards industrial or agricultural wastes to be used as raw material, and, then, lessen the cost of production [3]. Besides, the use of these precursors reduces residue generation in both rural and urban areas. The objective of this research was to produce low cost AC from vegetable waste materials. *Opuntia ficus indica* and *Casimiroa edulis* carbons were used to remove different types of textile dyes. Their adsorptive properties were then analyzed to find the relationship between their different structures and chemical surfaces.

2. Experimental

Materials. The six ACs were carbonized in a Thermolyne 48000 muffle furnace under their combustion gases atmosphere. Three were prepared from *Opuntia ficus indica* fruit waste and were designated: TunaAsh, CarTunaT, and CarTunaQ. The three adsorbents prepared from the *Casimiroa edulis* fruit waste were: CenZAP, CarZAPT, and CarZAPQ. TunaAsh and CenZAP consist of ashes obtained from sudden heating at 550 °C for 3 h; CarTunaT and CarZAPT consist of the materials carbonized at 400 °C for 3 h with controlled heating, the heating rates were 4.7 and 3.8 °C/min respectively; lastly, CarTunaQ and CarZAPQ consist of chemically activated carbons impregnated with H_3PO_4 and pyrolyzed at 400 °C during 3 h, the heating rates were 4.3 and 2.1 °C/min respectively. Only the chemically activated materials were washed with distilled water, to eliminate the H_3PO_4 excess, until a neutral pH was obtained in the washed solution. All materials were ground and sieved to obtain a particle size ranging from 0.25 to 0.84 mm.
To explore the effects of the carbonization process over the adsorptive properties of the ACs, different types of dyes, according to the Colour Index (CI) [4] classification, were selected. Table 1 shows the selected dyes and some of their properties. With the exception of vat dyes (water insoluble) and basic dyes, which are cationic molecules. The rest of the molecules of the dyes can be considered anionic. Three of the dyes were studied as model aqueous solutions and seven were found in wastewater produced from the dyeing process of cotton fabrics.

Table 1. Selected dyes and their characteristics

Dye (CI Generic Name)	Key	Textile Classification	λ_{max} (nm)	Solution type	pH	C_i (ppm)
Methylene Blue (Basic Blue 9)	BB9	Basic	665	Aqueous	6.1	50
Crystal Violet (Basic Violet 3)	BV3	Basic	590	Aqueous	6.0	50
Carminic Acid (Natural Red 4)	NR4	Acid	500	Aqueous	3.2	50
Sirius Green (Direct Green 1)	DG1	Direct	610	Wastewater	9.8	20
Sirius Turquoise (Direct Turquoise 86)	DT86	Direct	620	Wastewater	9.9	68
Vat Navy Blue	VNB	Vat	560	Wastewater	12	—
Vat Dark Grey	VDGY	Vat	400	Wastewater	12	—
Reactive Blue	RB	Reactive	610	Wastewater	11	—
Reactive Red	RR	Reactive	515	Wastewater	11	—
Reactive Orange	RO	Reactive	495	Wastewater	11	—

Methods. Static adsorption at 30 °C on every AC was studied using an adsorbent mass-solution volume (m/V) ratio of 0.01 g/L. The contact time was 24 h, although, in some cases, it was observed that within the first hour of contact, the dye had been completely removed from the solution. The adsorption results were compared with a commercial powder charcoal (Sigma de México) with a particle size ranging from 74 to 150 μm.
The quantification of the adsorbed dyes was done by spectrophotometry in the visible region using a HACH DR 2500 spectrophotometer at maximum wavelength absorbance. For BB9, BV3, NR4, DG1, and DT86, their respective calibration curves were constructed and used for quantification. For the rest of the dyes present in wastewater, the quantification was obtained by observing the decrease in the intensity of the main color band.
Fourier transform infrared (FTIR) spectra for the ACs were acquired using a PerkinElmer Spectrum One spectrometer with attenuated total reflectance (ATR) in the 4000 to 650 cm^{-1} spectral range.
The microscopic surface features of AC materials were observed with a JEOL model JSM 6300 scanning electron microscope operated a 30 kV, previously the samples were coated with gold.

3. Results and discussion

The organic functional groups of the activated carbons were determined by means of FTIR spectra (Figure 1). For CarTunaT and TunaAsh (Fig. 1b), the spectra are similar. Only a slight diminishing of the intensity of certain bands is observed. However, for CarTunaQ, its functional groups do not coincide with the previous ones. Both TunaAsh and CarTunaT exhibit bands at 1400, 1100, and 1050 cm^{-1} which are characteristically found in

IOP Conf. Series: Materials Science and Engineering **45** (2013) 012023 doi:10.1088/1757-899X/45/1/012023

O-H deformation and C-O stretching vibrations in primary, secondary, and tertiary alcohols and phenols. CarTunaQ shows a band at approximately 3500 cm^{-1} which indicates the presence of primary and secondary amines because of the N-H stretching vibrations. This is confirmed by the band found at 1600 cm^{-1} which indicates N-H deformation vibration for the same kind of amines.

CarTunaQ also presents a broad band at 1200 cm^{-1} because of the O-H deformation and C-O stretching vibrations for phenols. Furthermore, a band at 999 and a shoulder at 1070 cm^{-1}, characteristically found in phosphate groups [5], appear because of the H_3PO_4 impregnation treatment. In CarZAPQ (Fig. 1a), the appearance of bands at 999 and 1070 cm^{-1} characteristic of phosphate groups is also observed.

The carbons obtained from *Casimiroa edullis* fruit waste exhibit (Fig. 1a) a greater amount of functional groups, including N-H deformation (1500 cm^{-1}), O-H deformation and C-O stretching (1400, 1100, and 1050 cm^{-1}), and C-H stretching (alquenes at 3000 cm^{-1}) vibrations. In a similar way as *Opuntia ficus indica* waste carbons, ash (CenZAP) and thermal carbon (CarZAPT) exhibit practically the same bands.

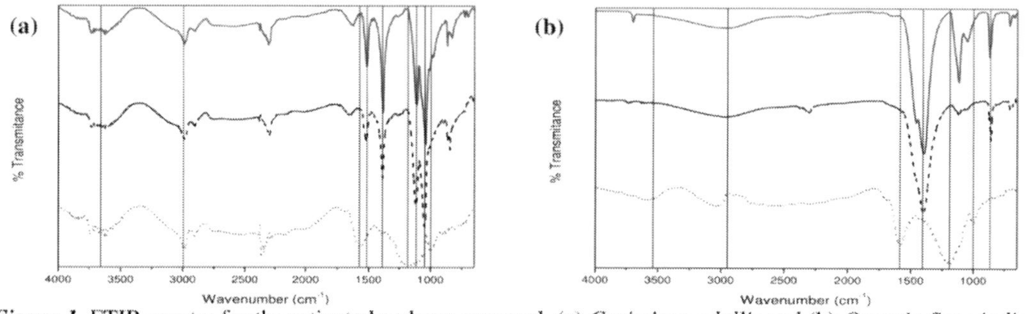

Figure 1. FTIR spectra for the activated carbons prepared. (a) *Casimiroa edullis* and (b) *Opuntia ficus indica* carbons. — Ashes, --- Physical, and ····· Chemical carbons.

Regarding their superficial appearance, the micrographs obtained by SEM (Figure 2) for ashes and physically activated carbons (a, b, d, e) obtained from both residues show the formation of irregularly shaped and sized cavities because of the irregular structure of the raw materials and the release of volatile compounds during the thermal treatment as well as small particles deposited on the surface.

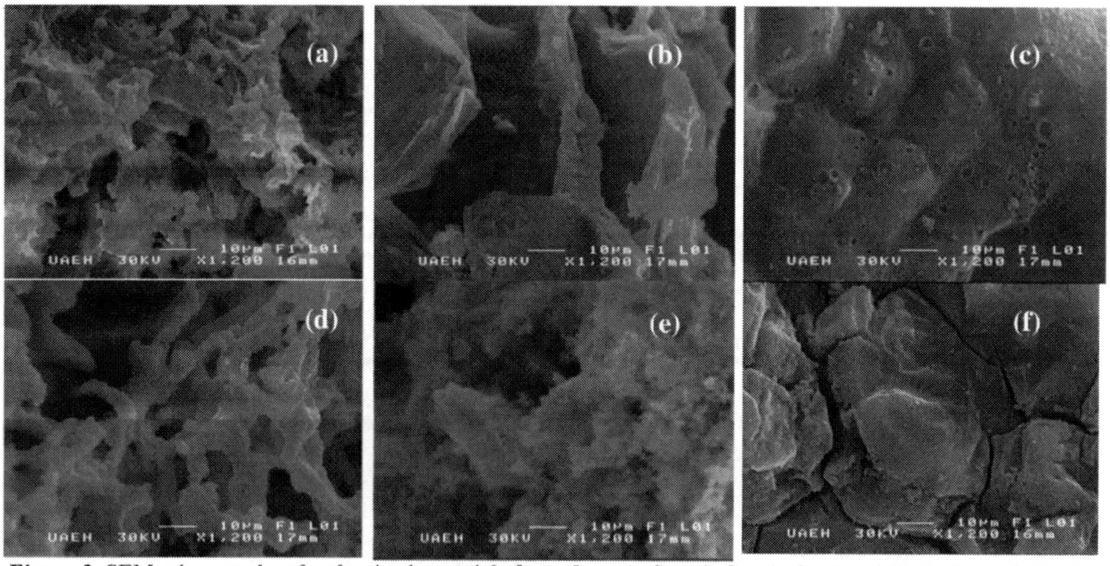

Figure 2. SEM micrographs of carbonized materials from *Opuntia ficus indica* (a, b, c) and *Casimiroa edulis* (d, e, f) wastes. Ashes (a and d), Physical activated carbons (b and e), and Chemical activated carbons (c and f).

Again, the carbons activated with H_3PO_4 turn out to be superficially different. CarTunaQ (c) shows superficial holes, and CarZAPQ (f) exhibits superficial cracking rather than a development in porosity as such.
The adsorption capability of ACs is expressed as the dye removal percentage, and it is presented in Figure 3. Graph (a) shows that acid, basic, and direct dyes are being retained excellently in the carbons, almost as it would be the case for a commercial carbon. Graph (b) shows the removal of vat and reactive dyes in wastewater, which turned out to be acceptable, but not as good as in the previous cases. This happens because wastewater is a very complex matrix that contains diverse compounds (salts like sodium carbonate, hydroxide, hydrosulfite or sulfate, and surfactants) used during the dyeing process that surely compete for active sites, thus inhibiting the interaction of the dyes.
The high adsorption of cationic and anionic dyes is confirmed with the IR spectrums which show the presence of anionic and cationic functional groups. These give a certain polarity to the carbons and, thus, the ability to interact with the ionic dyes. Furthermore, the micrographs show a great quantity of cavities in the carbons, which favor the development of the specific surface area and, hence, a better adsorption capacity.

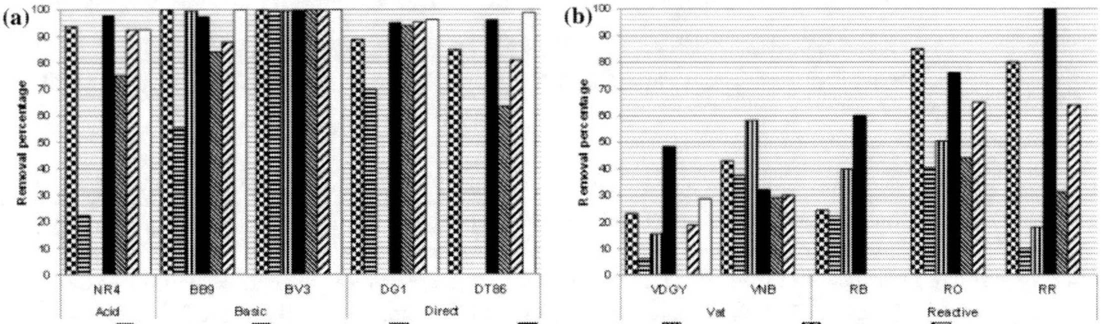

Figure 3. ▩ CarZAPQ, ▤ CarZAPT, ▥ CenZAP, ■ CarTunaQ, ▩ CarTunaT, ▨ TunaAsh, □ Sigma Carbon

4. Conclusions

Both *Opuntia ficus indica* and *Casimiroa edullis* fruit wastes in their carbonized forms are effective in the removal of textile dyes, reaching removal rates superior even to commercial AC, regardless of the smaller particle size this last one has. Hence, it is possible to produce activated carbons from these two residues which are common and abundant in the region (Puebla and Hidalgo states of Mexico). The adsorptive capabilities of the ACs depend strongly of their superficial functional groups and their morphology.

Acknowledgements. We thank Vicerrectoría de Investigación y Estudios de Posgrado, BUAP for financial support in the project PECA-ING11-G. Special thanks to H. Welti-Peláez for his review of this article's English language. Thanks to R. Radillo-Ruíz for wastewaters supply.

5. References

[1] Peláez-Cid AA and Teutli-León MMM 2012 *Lignocellulosic Precursors used in the Synthesis of Activated Carbon - Characterization Techniques and Applications in the Wastewater Treatment*, Edited by Hernández-Montoya V and Bonilla-Petriciolet A (Croatia: InTech) Chapter 1 pp. 1-18.
[2] Elizalde-González MP, Mattusch J, Peláez-Cid AA, and Wennrich R 2007 Characterization of adsorbent materials prepared from avocado kernel seeds: Natural, activated and carbonized forms, *J. Anal. Appl. Pyrolysis* **78** 185–193.
[3] Elizalde-González MP and Hernández-Montoya V 2007 Characterization of mango pit as raw material in the preparation of activated carbon for wastewater treatment, *Biochem. Eng. J.* **36** 230–238.
[4] Colour Index International, Fourth Edition Online, Society of Dyers and Colourists & American Association of Textile Chemists and Colorists, www.colour-index.com.
[5] Suárez-García F, Martínez-Alonso A, and Tascón JMD 2004 Activated carbon fibers from Nomex by chemical activation with phosphoric acid, *Carbon* **42** 1419-1426.

IOP Conf. Series: Materials Science and Engineering **45** (2013) 012024 doi:10.1088/1757-899X/45/1/012024

Influence of nitrogen in TiAlNO coatings fabricated by co-sputtering

García González L.*[1], Hernández Torres J.[1], Flores Ramírez N.[2], Argumedo Moreno P.[3], López Velázquez A.[4], Araujo López D.J[5]. & Courrech Arias A.M.[1]

[1] Centro de Investigación en Micro y Nanotecnología, Universidad Veracruzana, Veracruz, México

[2]División de Estudios de Posgrado de la Facultad de Ingeniería en Tecnología de la Madera, Universidad Michoacana de San Nicolás de Hidalgo, Michoacán, México

[3]Instituto Tecnológico Superior de Poza Rica, Veracruz, México

[4]Facultad de Ingeniería Mecánica Eléctrica, Universidad Veracruzana, Veracruz, México

[5]Facultad de Ingeniería Química, Universidad Veracruzana, Veracruz, México

*leagarcia@uv.mx

Abstract. TiAlNO coatings were obtained on corning glass by means of the co-sputtering technique. Two series were fabricated. In the first one the argon flux was maintained at 12 sccm, while the nitrogen flux was varied in 3, 6 and 9 sccm; in the second one the argon flux was 15 sccm and the nitrogen flux was varied in 2, 3, 4, 5 and 6 sccm, which produced a working pressure variation. For co-sputtering in the first series, three targets were occupied: titanium, alumina and aluminium; meanwhile in the second series, two targets were used: titanium monoxide and aluminium plus an amount of titanium, specifically, 3.5 gr in the aluminium target, occupying 40% of the total effective area of the target. Structural, mechanical and electrical properties were obtained through X-ray diffraction, the Vickers microhardness test and resistivity probes, respectively. For first series, X-ray diffraction detected titanium nitride (TiN) in the crystalline phase; for the second series, aluminium nitride and titanium oxide $(TiO)_{3.38}$. The difference in the crystalline structure obtained for the series explains the microhardness values; in the first series a maximum value of 20.4 GPa at 2 sccm of nitrogen was obtained, whereas in the second series the maximum value obtained was 8 GPa. As one can observe, the increase in nitrogen flux is not an advantage to hardness, but in electrical resistivity it has a positive effect. The results suggest that TiAlNO is versatile material that could be used for both mechanical and electronic applications; however, it is important to control fabrication conditions.

1. Introduction

Nowadays there is a need for materials with excellent and diverse properties to be applied in different scientific areas like metal mechanics and the electronic industry. Metal transition coatings fabricated by physical methods are used for mechanical, optical and electrical applications due to their excellent thermal and chemical stability[1], good wear resistance, attractive colors and excellent diffusion barrier properties[2]. The most used synthesized coatings are TiN[3] and CrN[3]; however, these coatings have been developed by adding another element like Al, Si, C, Zr, or even by being fabricated in an oxidant environment to obtain oxides[4] (ceramics materials); said addition improves the properties of the material base but depends on property requirements.

H.C. Kim and T.L. Alford[5] studied the behaviour of $TiAl_xN_yO_z$ coatings as barriers against diffusion, using the sputtering technique. For this they used a TiAl target and incorporated gases of Ar, N_2 and O_2. They found the maximum value to be 4.2 μΩ-cm, and that this value diminishes as the temperature from the thermic treatment increases.

J. Sjölén and co-authors[6] analyzed the structure and the mechanical properties of TiAlNO coatings fabricated with an electric arc. They found hardness values up to 35 GPa, which diminished as the oxygen flow was increased. Such high values of hardness, measured by means of nano-indentation techniques, are attributed to a nano-crystalline phase of TiN immersed in amorphous phases of Al_2O_3, Ti(O,N), and AlN. Similarly, Fanghua Mei and co-authors[1] obtained 35 GPa of hardness by using a mosaic type of Ti/Al target for sputtering, varying the gases of oxygen and nitrogen. Other authors have focused on studying the chemical stability diagram of this quaternary system[7].

In this work, TiAlNO coatings were fabricated by co-sputtering under different working conditions; this resulted in two different series of samples showing a difference between each other. The structure, hardness and resistivity of both series were analyzed. In the first series, high hardness and low resistivity values were found, whereas high resistivity and low hardness values were found in the second series.

2. Experimental Details

The two series of coatings were fabricated in a Sputtering V3 INTERCOVAMEX system. In the first series, we used three targets: Titanium-Ti (99.99%) in DC source, alumina-Al_2O_3 (99.99%) in RF source and silicon-Si (99.99%) in RF source, maintaining Argon gas at 12 sccm and nitrogen flux variation at 3, 6 and 9 sccm. In the second series, we used two targets: titanium monoxide-TiO (99.99%) in DC source and aluminium-Al (99.99%) plus 3.5 gr of Ti in RF source, using Argon flux at 15 sccm and 2, 3, 4, 5 and 6 sccm of nitrogen flux variation. Table 1 describes specific details of coatings series fabrication.

SERIES	SAMPLES	N_2 Flux (sccm)	Ar Flux (sccm)	TARGETS	DC POWER (W)	RF POWER (W)	RF POWER (W)	Thickness (μm)
1	S1	3	12	Ti, Al & Al_2O_3	200	150	200	0.5
	S2	6						
	S3	9						
2	SA	2	15	TiO & Al+ 3.5 gr Ti	100	250	N/U*	0.5
	SB	3						
	SC	4						
	SD	5						
	SE	6						

Table 1. Fabrication Details

3. Characterization

3.1. X-ray diffraction

For X-ray analysis an X-ray diffractometer BRUKER D8 Advance with Göbbel mirror primary optics and scintillation detector was used, along with generator parameters of 40 kV and 40 mA for both sets and with θ=2° and 2θ= 30-90°. The X-ray diffraction data of the first series in Figure 1 show only the TiN crystalline phase (PDF: 01-087-0632). In the case of the second series, the aluminium nitride (AlN) (PDF: 00-025-1495) and titanium oxide $(TiO)_{3.38}$ crystalline phases (PDF: 01-075-0310) were observed, Figure 2.

3.2. Vickers' Microhardness

A Vickers' Microhardness Test Machine (HM-124, MITUTOYO) was used to obtain the mechanical properties of the coatings. The applied loads were 0.001 kgf up to 0.5 kgf during 10 seconds, each one being applied three times. The hardness values of the coatings were obtained by applying Koursunky's model Eq.1[8]. For the first series the maximum value was 20.4 GPa for S1 and the minimum value was 6.52 GPa for S2. For the second series, sample A (SA at 2 sccm N2) was the hardest with 7.25 GPa,

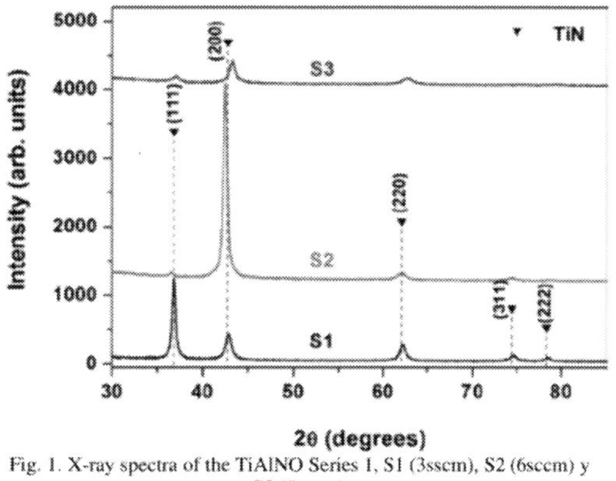

Fig. 1. X-ray spectra of the TiAlNO Series 1, S1 (3sscm), S2 (6sccm) y S3 (9sccm).

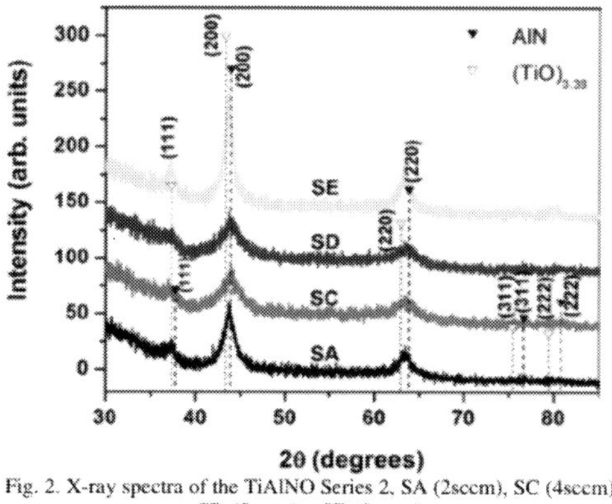

Fig. 2. X-ray spectra of the TiAlNO Series 2, SA (2sccm), SC (4sccm), SD (5sccm) y SE (6sccm).

whereas sample C (SC at 4 sccm) was the least hard with 4.85 GPa. A comparison between the series of coatings is shown in Figure 3, where the difference is distinguishable, the value of the least hard sample in Series 1 being almost the same as the maximum value in Series 2.

$$H_{TOT} = H_S + \frac{H_f - H_S}{1+k\beta^2} \qquad \text{Eq. 1}$$

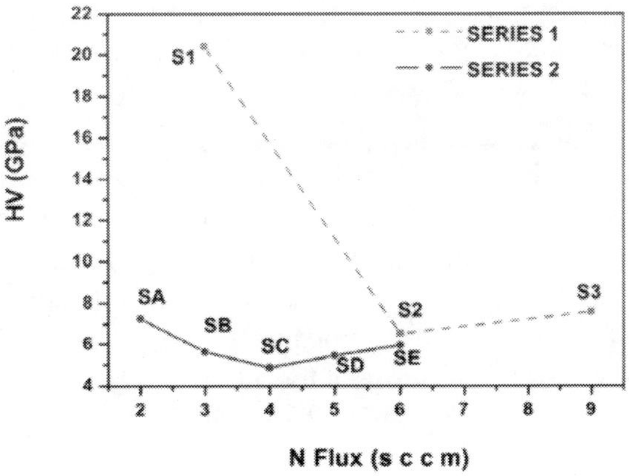

Fig. 3. Hardness vs. N_2 Variation, comparison of Series 1 and Series 2.

3.3. Electrical Resistivity

The electrical measurements of the series were obtained with a Resistivity Test Fixture, KEITHLEY 6517-B, 8009. In series 1, all the samples have a semiconductor/insulator behavioural trend with values between 3.88×10^3 Ω-cm and 4.081×10^5 Ω-cm; on the other hand, series 2 presented insulator behaviour, the values being between 5.77×10^6 Ω-cm and 9.904×10^2 Ω-cm.

4. Results and Discussion

In figure 1, it is observed for the first series that all the diffraction peaks correspond to the TiN crystalline phase. Because there is no evidence of any other crystalline phase in Figure 1, energy dispersive spectroscopy measurements were realized, but only two elements, titanium and nitrogen, were detected. In this case, it was assumed that aluminium and oxygen could not be deposited in the coatings under the initial growth conditions proposed. The nitrogen flow variation causes a change in the preferential crystallographic direction of the coating S1 to S2, which is maintained up to the S3 coating and can be noted as an intensified diffraction peak. Moreover, upon increasing the nitrogen flow it can observed that the diffraction peak located at $2\theta = 42.8°$ shows a shift to the right with respect to the reference line, which indicates that there are compressive stresses in the coatings. On the other hand, the increase in the nitrogen flow considerably affects the hardness values, decreasing them from 20 GPa to 6.5 GPa. The growth parameters of the first series of coatings were changed without obtaining the values of resistivity, because it was not possible to obtain the quaternary system of TiAlNO.

In the second series, TiAlNO coatings with a thickness of 500 nm were achieved. With XRD, two crystalline phases were identified, AlN and TiO. Upon increasing the nitrogen flow, the diffraction peaks showed a shift toward the left with respect to the reference line, indicative of the presence of tensile stress. At high values of nitrogen flow, the shift and the intensity of the diffraction peak are greater. Energy dispersive spectroscopy measurements confirmed the presence of the 4 elements of the quaternary system, Al, N, O and Ti, Figure 4. The harder coating was obtained at 2 sccm and corresponds to the SA coating with a value of 8 GPa; when the value of the nitrogen flow increases, the hardness decreases down to a minimum value of 4.8 GPa for the SC coating. To continue increasing the nitrogen flow, the hardness values are increased again until reaching a value of 6 GPa for the SE coating. Electrical resistivity values decrease with the increase in the hardness values, reaching a maximum value of 5.77×10^6 Ω-cm for the SC coating; therefore the flow of nitrogen is an important variable to consider in deposition processes, depending on the application required. Growing conditions for the SC coating would favor its application in the diffusion barriers in integrated circuits[5], while the SA coating conditions have greater application in the metalworking industry, where high-speed steel parts are used for cutting[9-11].

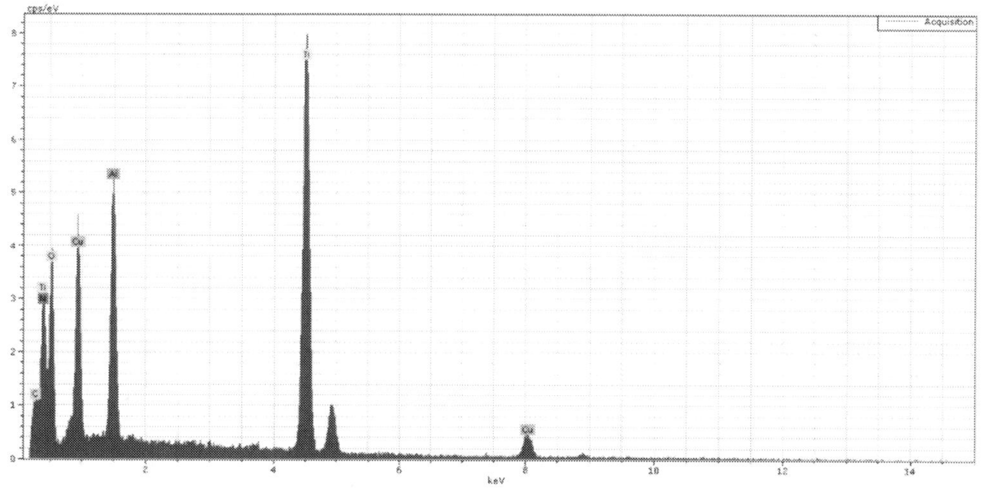

Fig. 4. Results of EDS confirming the presence of the 4 elements desired.

References

[1] Mei F, Dong Y, Li Y and Li G 2006 *Mater. Lett.* **60** 375
[2] Kim H C and Alford T L 2004 *Thin Solid Films* **449** 6
[3] Meneve J, Vercammen K, Dekempeneer E and Smeets J 1997 *Surf. Coat. Tech.* **94-95** 476
[4] Hu Y A, Ma T B 2011 *Compr. Nanosci. Technol.* **3** 383
[5] Kim H C and Alford T L 2004 *Thin Solid Films* **449** 6
[6] Sjölén J, Karlsson L, Braun S, Murdey R, Hörling A and Hultman L 2007 *Surf. Coat. Tech.* **201** 6392
[7] Seifert H J, Kussmaul A and Aldinger F 2001 *J. Alloy. Compd.* **317–318** 19
[8] Korsunsky A M, McGurk M R, Bull S J and Page T F 1998 *Surf. Coat. Tech.* **99** 171
[9] Mitterer C *et al* 2003 *Surf. Coat. Tech.* **163–164** 716
[10] Podgornik B, Zajec B, Bay N and Vižintin J 2011 *Wear* **270** 850
[11] Budke E, Krempel-Hesse J, Maidhof H and Schüssler H 1999 *Surf. Coat. Tech.* **112** 108

Effects of temperature on the morphology and optical properties of ZnS thin films deposited by chemical bath

P.E. Martín-Várguez, O. Ceh, I. J. González-Panzo, S. Tec-Yam, R. Patiño, and A. I. Oliva
Centro de Investigación y de Estudios Avanzados del IPN, Unidad Mérida. Departamento de Física Aplicada. Km 6 Antigua Carretera a Progreso, A.P. 73-Cordemex, 97310 Mérida Yucatán

E-mail: oceh@mda.cinvestav.mx

Abstract. Zinc sulphide thin films were deposited on Corning glass substrates by the chemical bath deposition technique at different temperatures. The influence of the bath temperature and deposition time on the morphological and optical properties of the ZnS films are herein investigated. ZnS films were deposited by changing the bath-temperature from 50 °C to 90 °C, and deposition times from 60 to 160 min. Thin and transparent films were obtained with thicknesses from 10 to 90 nm with the increment of the bath temperature, meanwhile the band gap energy E_g values diminishes from 4.15 to 3.4 eV. The quality of the ZnS film surfaces was also influenced by increasing the bath temperature, as showed by the reduced grain size and the increase of roughness, obtained from atomic force microscopy images. ZnS films of good optical quality were obtained at 90 °C with a mean value of E_g=3.56± 0.03 eV.

1. Introduction

Due to its suitable optical properties, zinc sulphide (ZnS) emerges as an attractive semiconductor material for electronic and solar applications. ZnS thin films are commonly used as optical filter due to its appropriate band gap energy value of E_g = 3.68 eV at room temperature [1-3]. They can also be used as buffer layer or as antireflective coating [4] for improving the efficiency of thin film solar cells. Efficiencies above 18 % have been reported for ZnS/Cu(In,Ga)Se_2 (CIGS) thin film solar cells [5]. Cadmium sulphide (CdS) has been typically used as buffer layer on CIGS thin solar cells to protect the Cu(In,Ga)Se_2 junction from possible damage during the zinc oxide (ZnO) sputtering deposition, which would cause unwanted modifications to the surface of the CIGS absorber. The efficiency of the CdS/CIGS solar cells drops at short wavelengths due to optical absorption losses from the CdS layer. The solar absorption can be improved by replacing the CdS film with other semiconducting material with a wider energy band gap. Because of its appropriate optical characteristics, ZnS arises as a good candidate for being used as a buffer layer in solar cell applications. The ZnS higher band gap energy (3.68 eV) [6] compared to that one of the CdS (2.42 eV) makes the ZnS wavelength range of transparency wider (above 330 nm) than the CdS one (above 520 nm) [8], increasing then the amount of sunlight on the photovoltaic conversion process. In addition, replacing CdS with ZnS is environmentally attractive by its non-toxicity. The effects of the chemical bath temperature on the ZnS rate of growth and of annealing temperature on the structural properties ZnS films have been studied [9,10]. In this work, ZnS thin films were deposited on glass substrates by the chemical bath deposition (CBD) technique varying the bath temperature and the deposition time, in order to study the influence of such parameters on the ZnS morphological and their optical properties.

2. Theory

The solubility product value of ZnS at 25 °C has been reported as $K_{sp} = 10^{-24.7}$ [11]. As a consequence, ZnS precipitation may take place even at very low ion concentrations of Zn^{+2} and S^{-2}. Controlling the free Zn^{+2} concentration, becomes an important issue in order to achieve such low-concentration precipitations. The use of a suitable complexing agent, promotes the releasing of low concentrations of Zn ions. In this work NH_3 is proposed as the complexing agent. The following chemical reactions are proposed as the mechanism for the formation of ZnS films:

$$NH_4^+ + OH^- \leftrightarrow NH_3 + H_2O \quad (1)$$
$$Zn^{2+} + 4NH_3 \leftrightarrow [Zn(NH_3)_4]^{2+} \quad (2)$$
$$SC(NH_2)_2 + 2OH^- \leftrightarrow S^{2-} + H_2NCN + 2H_2O \quad (3)$$
$$[Zn(NH_3)_4]^{2+} + S^{2-} \rightarrow ZnS\,(s) + 4NH_3 \quad (4)$$

Given that, the ZnS present a direct band gap energy E_g, its value can be estimated from the relation:

$$\alpha^2(h\nu) = A(h\nu - E_g) \quad (5)$$

where α is the absorption coefficient, h the Planck´s constant, ν the incident light frequency, and A is a constant which depends on the type of material. Thus, the E_g value can be estimated as the $h\nu$ value obtained at the intersection with the x-axis ($\alpha^2=0$) from the α^2 vs $h\nu$ plot.

3. Experimental

ZnS thin films were deposited on Corning glass substrates (10 x 25 mm^2). Substrates were sequentially cleaned with soap, trichloroethylene, acetone, isopropyl alcohol, and de-ionized water, and dried with air as a last step. Teflon sample-holders were used to vertically support the substrates within the bath. The chemical bath was composed by the following analytical grade reagents: 80 ml of 0.02 M $ZnCl_2$, 80 ml of 0.20 M $SC(NH_2)_2$, 80 ml of 1.50 M NH_4NO_3, and 200 ml of 0.90 M KOH. A glass beaker was used as a container for the films deposition. In order to control the bath temperature during the heating process, a hot plate with an adapted proportional-integral-differential (PID) feedback system was used, with variations of ± 0.5 °C. Five groups of ZnS samples at different temperature (each) were obtained by changing the bath temperature from 50 °C to 90 °C in steps of 10 °C. For each bath temperature, five samples were deposited with different deposition time ranging from 60 min to 140 min. For the chemical bath preparation, $ZnCl_2$, KOH, and NH_4NO_3 reagents were subsequently poured (in this order) to the glass beaker and magnetically stirred and heated until the work temperature was reached. Once such a temperature was reached, the solution of $SC(NH_2)_2$ was finally added to the container, and the ZnS deposition process took place. During the deposition process, a nearly constant pH~11.9 was measured. After the defined-deposition time has passed, each sample was taken out from the bath, and ultrasonically cleaned to eliminate surface residues. Deposited ZnS films were optically characterized with a StellarNet EPP-2000 spectrophotometer to measure the band gap energy. Atomic force microscopy (AFM) images were obtained with an AMBIOS Universal SPM equipment to analyse the surface morphology, and thickness was measured by the Dektak 8 profilometer.

4. Results

The ZnS films deposited by CBD on glass substrates presented low thickness values (10 to 90 nm), good transparency, and good adherence to the substrates. Figure 1 shows a group of representative AFM images of ZnS films deposited at 90 °C and different deposition times. Small differences on the surface morphology can be observed. The rms-roughness and grain size values estimated from the AFM images as a function of deposition time and bath temperature are showed in Figures 2 and 3. From Figure 2, it is observed that the roughness value increases with the increase of both deposition time and bath temperature. The grain size do not show a very clear tendency with deposition time (Figure 3), but at high temperatures, seems to diminish the average values of grain size, as observed with AFM. Higher roughness and grain size values are commonly associated with the increment of film thickness (i.e. deposition time) until a saturation region is reached. However, some chemical compounds different to ZnS could co-precipitate at low temperatures.

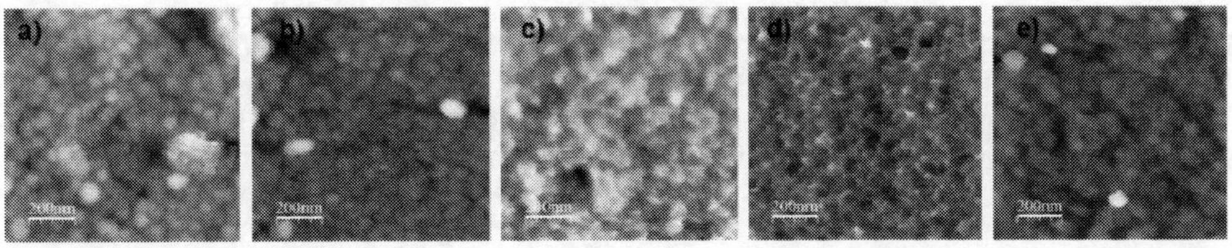

Figure 1. AFM images (1x1 μm^2) of ZnS/glass films deposited at 90 ºC: a) 60 min, b) 75 min, c) 90 min, d) 105 min, and e) 120 min, as deposition time.

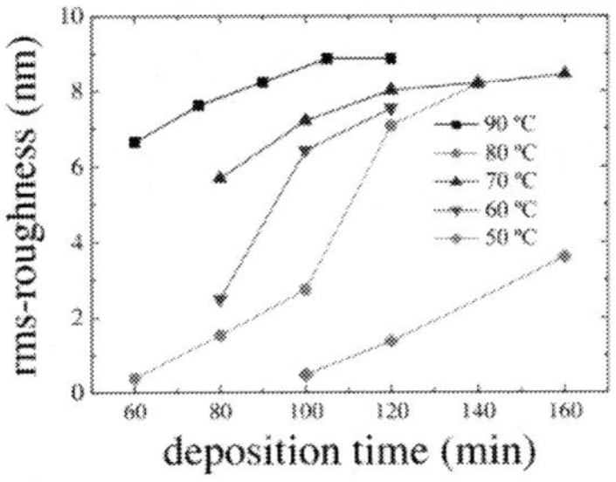

Figure 2. Roughness value vs deposition time of ZnS films deposited at different bath temperatures.

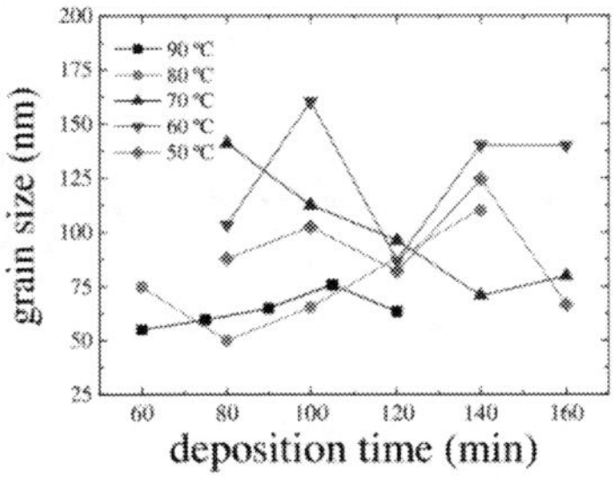

Figure 3. Grain size vs deposition time of ZnS films deposited at different bath temperatures.

Figure 4 shows the α^2 *versus* energy plot of the five samples of ZnS films deposited at 90 ºC and different deposition times. An increase of the absorbance coefficient with the increase of deposition time can be clearly observed which is directly related with the thickness of the films. A mean value of the band gap energy $E_g = (3.56 \pm 0.03)$ eV was estimated for all the films deposited at 90 ºC, a slightly lower value than the 3.68 eV value reported in the literature [12].

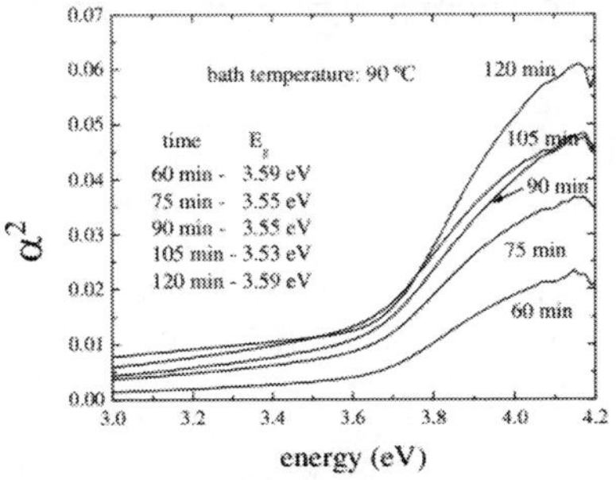

Figure 4. α^2 vs energy plot for ZnS films deposited at 90 ºC and different deposition times. The band gap energy value is reported for each case. A sequential formation of the absorption edge with deposition time is observed.

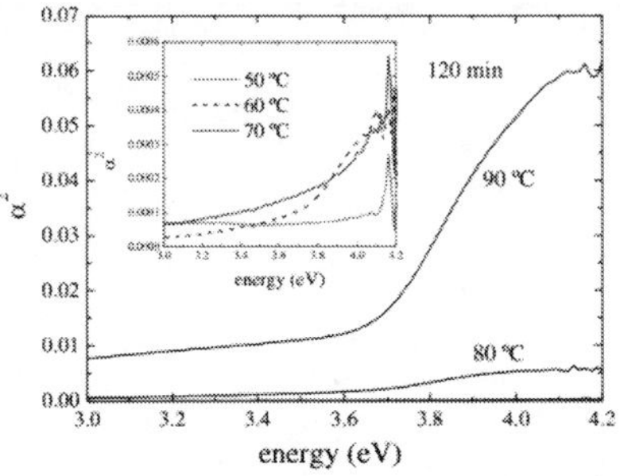

Figure 5. α^2 vs energy plot for ZnS films deposited during 140 min at different temperatures. The formation of the absorption edge can be appreciated by increasing the bath temperature.

This difference could be associated with defects both on the film stoichiometry and on the crystalline structure during deposition. Figure 5 shows the α^2 vs energy plot of the ZnS films deposited during 120 min at different temperatures. Inset in Figure 5 corresponds to the thinnest ZnS films deposited at lower temperatures which presented the lowest absorbance values. An important increase in absorbance intensity is observed with the increase of the bath temperature. A formation of the absorption edge of the ZnS films is observed when the bath temperature increases. Thus, the band gap energy value can be clearly determined from this edge. By increasing the bath temperature, the chemical reaction is accelerated, which makes lower the time deposition required to obtain determined thickness due to the rapid exhaustion of the chemical bath. Thus, the bath temperature strongly influences on the obtained optical properties of the ZnS films, as can be concluded from the presented results. Profilometry measurements were carried out on the ZnS films through an intentionally formed step with a HCl-10% solution on a corner. Reported thickness values varied from 10 to 90 nm, being the higher values for the films deposited at higher temperatures. Experimental data for ZnS films properties obtained for different temperatures and 120 min as deposition time, are reported in Table I.

Table I. Experimental values obtained from ZnS films deposited at different temperatures and 120 min as deposition time.

Deposition temperature (°C)	Grain size (nm)	Roughness rms (nm)	Band gap energy (eV)	Thickness (nm)
50	82	1.4	4.11	11
60	87	7.5	3.80	12
70	96	8.0	3.62	24
80	98	7.1	3.53	31
90	64	8.9	3.59	86

5. Conclusions

ZnS thin films were deposited on glass substrates by chemical bath deposition at bath temperatures ranging from 50 to 90 °C and deposition times ranging from 60 to 160 min. Better ZnS films, thicker transparent, and with a band gap energy value close to ZnS, were obtained for films deposited at 90 °C as bath temperature. A mean value of $E_g = 3.56 \pm 0.03$ eV was obtained for the group of these better ZnS films. The temperature of the bath was found to have a strong influence on the surface quality (roughness), the grain size, and the gap energy obtained.

References

[1] Ben T, Kamoun N, Kanzari M, and Bennaceur R 2006 *Thin Solid Films* **500** 4
[2] Ben T, Kamoun N, and Guasch C 2008 *Appl. Surf. Sci.* **254** 5039
[3] Ge Y H, Shi W M, Xu J, Y H Qiu and. Wei GP 2005 Proc. 15th International Photovoltaic Science and Engineering Conference Shanghai, China October 10-15 p. 262
[4] Gangopadhyay U, Kim K, Dhungel S K, Saha H, and Yi J 2007 *Advances in OptoElectronics*, vol. 2007, Article ID 18619
[5] Contreras M A, Nakada T, Hongo M, Pudov A O, and Sites J R. 2003 Proc. 3rd World Conference of Photovoltaic Energy Conversion. Osaka Japan May 11-18 p. 570
[6] Göde F, Gümüs C, and Zor M 2007 *J. Cryst. Growth* **299** 136.
[7] Arenas O L, Nair M T S, and Nair P K. 1997 *Semicond. Sci. Technol.* **12** 1323
[8] Negami T, Aoygi T, and Satoh T 2002 Proc. 29th IEEE, PVSC, New Orleans, p. 656
[9] Liu Q, Guobing M, and Jianping A 2008 *Applied Surface Science* **254** 5711
[10] Roy P, Ota J R, and Srivastava S K 2006 *Thin Solid Films* **515** 1912
[11] Cotton F A, and Wilkinson G 1980 *Advanced Inorganic Chemistry* (Wiley-Interscience New York)
[12] Nadeem M Y, and Ahmed W 1999 *Turk J. Phys.* **24** 651

Polymer concrete reinforced with recycled-tire fibers: Mechanical properties

E Martínez-Cruz [1], G Martínez-Barrera [2]* and M Martínez-López [1]

[1] Posgrado en Ciencia de Materiales, Facultad de Química, Universidad Autónoma del Estado de México, Toluca México.

[2] Laboratorio de Investigación y Desarrollo de Materiales Avanzados (LIDMA), Facultad de Química, Universidad Autónoma del Estado de México, Km.12 de la carretera Toluca-Atlacomulco, San Cayetano 50200, México.

E-mail:gonzomartinez02@yahoo.com.mx

Abstract. Polymer Concrete was reinforced with recycled-tire fibers in order to improve the compressive and flexural strength. Polymer concrete specimens were prepared with 70% of silicious sand, 30% of polyester resin and various fiber concentrations (0.3, 0.6, 0.9 and 1.2 vol%). The results show increment of 50% in average of the compressive and flexural strength as well as on the deformation when adding 1.2 vol% of recycled-fibers.

1. Introduction

The development of polymer concrete (PC) has been of scientific and technological interest around the world since last three decades due to the versatility of its applications. There are several important features in PCs such as fast curing, light weight, good linkage with steel reinforcements, high compressive and flexural strength as well as durability that make it a suitable material for the construction industry [1-5]. PC is formed by a thermoset resin used as binder of natural or artificial aggregates; it is known that polymer concrete has mechanical properties higher than those for Portland cement concrete (PCC) but with a little drawbacks associated to brittle behavior, which can be solved adding fillers as fibers or particles.

The final properties of PCs rely on several criteria: polymer binder, mineral aggregates, type of reinforcement as well as the concentration of each component. In the polymer concrete elaboration is common the use of polyester resin by its low cost compared to other polymeric binders. It is not common the use of fiber reinforcements on polymer concrete like the one presented in the Portland cement concrete (PCC), thus a lack of information of such reinforcements is evident. Some fibers have been used for this purpose as steel, glass, carbon, nylon, polyester, propylene, among others; however in order to reduce the environmental impact of industrial or postconsumer waste, recycled fibers has been used. They offer advantages in reducing waste and conserving resources [6]. Recycled fibers can improve the mechanical properties of PCs. Wood wastes from timber industry are used as reinforcement of epoxy-polymer concrete and the flexural strength, fracture toughness and fracture energy are improved; similar behavior are observed for PCs with glass fiber as reinforcement.

The recycling and reutilization of cross-linked elastomers are difficult due their 3-D formed network, nevertheless it is necessary to found wise-strategies for reuse and to avoid ground contamination [7]. Modifications have been made to PCC with recycled-tire-rubber particles due to the benefits on the tensile and flexural strength of the mixtures. The natural and synthetic rubbers such as styrene-butadiene-styrene (SBS) and styrene-butadiene-rubber (SBR) are the raw material in the production of tires; the natural rubbers provide elastic properties while the synthetics provide thermal stability [8].

2. Experimental

2.1. Specimen manufacturing
Standard prismatic molds (40x40x160 mm) were used to formulate polymer concrete specimens according to CPT PC-2 recommendation. The polymer concrete specimens were elaborated with an isophthalic polyester resin (Aropol™ IS 4633) and a siliceous sand (SP55-Sibelco). The resin was accelerated by 1% of cobalt octoate, and the methyl ethyl ketone peroxide was used as initiator. The sand with granulometry uniform had an average diameter of 245 μm.

The polymer concrete specimens (taken as a control) consisted of 70 vol% of siliceous sand and 30 vol% of polyester resin. As reinforcements of polymer concretes were used recycled-tire rubber fibers at four different concentrations (0.3, 0.6, 0.9 and 1.2 vol%), these were taken into account in accordance to the sand concentrations. The recycled-fibers had a size distribution from 0.5 to 3.5 mm length. Before preparing specimens both aggregates, siliceous sand and fibers were dried in an oven for 3 hours at 60°C in order to guarantee the absence of water.

2.2. Mechanical tests
All specimens were cured for 24 hours at 30°C and post-cured at 80°C for three hours before being submitted to experimental testing. The mechanical tests of PC specimens were carried out in an Instron universal testing machine, with a load cell of 100 kN. The flexural testing was according to CPT PCM-8 standards at a rate of 1 mm/min for three-point bending tests. Load-displacement curves and the maximum load for the collapse bending were recorded. The compressive testing at a loading rate of 1.25 mm/min was done, following the UNE 83821:1992 test standard.

2.3. Morphological characterization
The surface of recycled-tire rubber fibers were observed with a scanning electronic microscope (SEM) in a JEOL model JSM-5200 machine, in the backscattering-electron mode.

3. Results and Discussion

3.1 Compressive and Flexural Strength
Compressive strength results of reinforced polymer concretes are shown in Figure 1a. It can be noticed two well-defined stages: a) In the first there is a diminution in compressive strength for polymer concretes with 0.3 and 0.6 vol% of recycled fibers, it means 54% of diminution respect to polymer concrete without fibers; b) Then the values increase for specimens with 1.2 vol% of recycled-fibers.

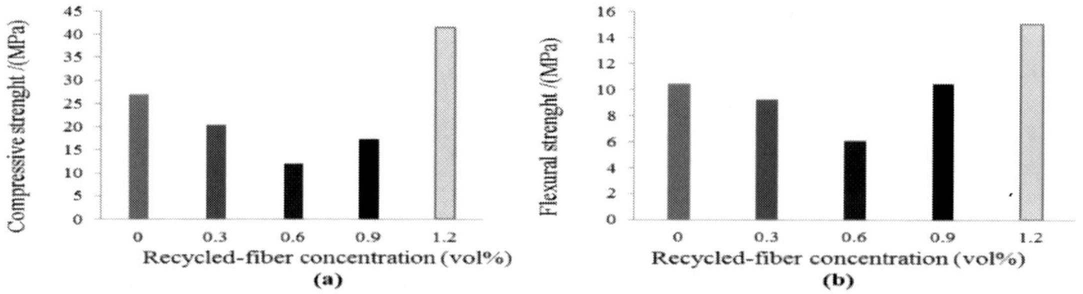

Figure 1. Polymer concrete with recycled-fiber: (a) Compressive strength, (b) Flexural strength.

With exception of specimens containing 1.2 vol% of fibers; all values are lower than those for polymer concrete without fibers. The highest value, namely 41 MPa means an improvement of 54% compared to polymer concrete without fibers.

The flexural strength behavior is similar when it is compared with compressive strength behaviour (see Figure 1b). The highest value, namely 15 MPa means an increase of 43%. We remind that the elastomeric fibers has a high Poisson ratio of approx. 0.5, this feature is noted when adding 1.2 vol% of fibers which is enough to "overcome" the poor interaction between tire-rubber fibers (an elastomer) and the polyester resin (matrix). A great quantity of different fiber sizes leads a greater distribution that can fill the interstitial spaces with smaller fibers.

3.1. Compressive Strain at Yield Point and Compression Modulus of Elasticity

As shown in Figures 2a and 3, a maximum compressive strain (20%) is obtained when 0.9 vol% of fibers are added, respect to plain polymer concrete, showing the influence of the elastomer on the strain behavior. A different behavior is observed for the elasticity modulus (see Figure 2b), which decreasing according to the fiber concentration increase. As we know a lower elasticity modulus implies a ductile material unlike a rigid material. In our case the lowest value is obtained for polymer concrete with 0.6% of fibers; which is 53% minor than polymer concrete without fibers, generating a more ductile material. An unexpected result is the increment of elasticity modulus for polymer concrete with 1.2% of recycled-fibers, which is 18% higher than that of polymer concrete without fibers.

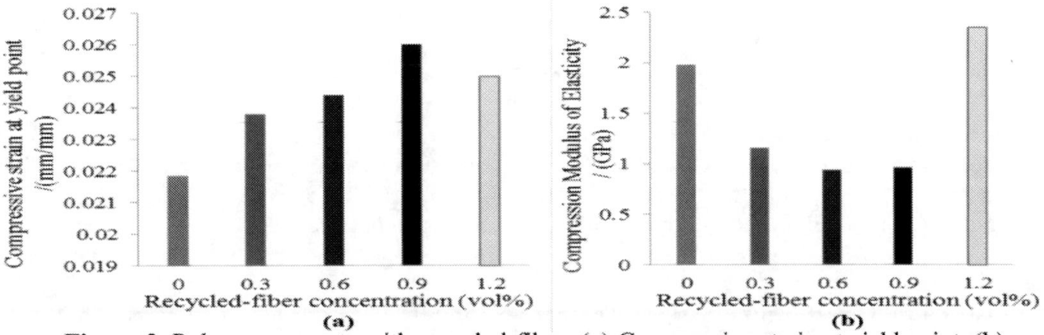

Figure 2. Polymer concrete with recycled-fiber: (a) Compressive strain at yield point, (b) Compression Modulus of Elasticity of polymer concretes.

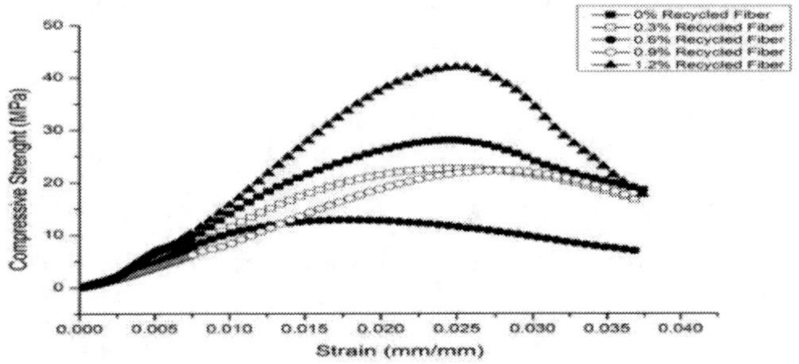

Figure 3. Compressive strength vs strain of polymer concrete with recycled-fiber.

In Figure 4 the fiber surface characteristics of the recycled-tire fibers are shown. Some fibers show roughness on their surface and others smooth surfaces. The average diameter of recycled fibers varies from 10 to 35 μm and the length from 500 to 3500 μm. The roughness surface can allow anchoring between the tire-rubber fibers (an elastomer) and the polyester resin (matrix). For the lowest concentrations a poor elastomer-matrix adherence is found, but when increases the volume fraction of fibers, mechanical interaction are augmented due to the higher amount of fibers, therefore improvements on the mechanical properties are obtained.

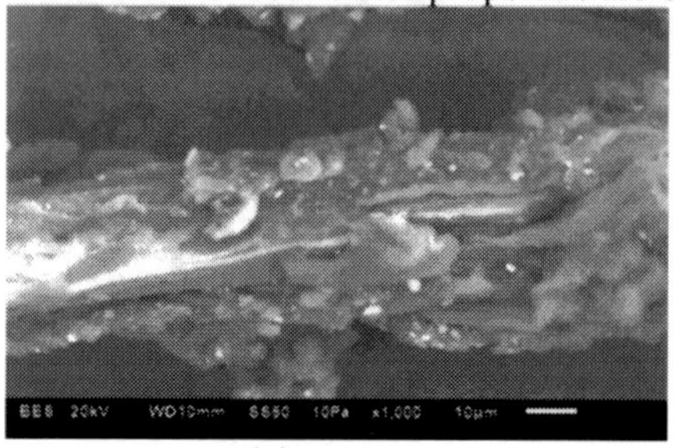

(a)

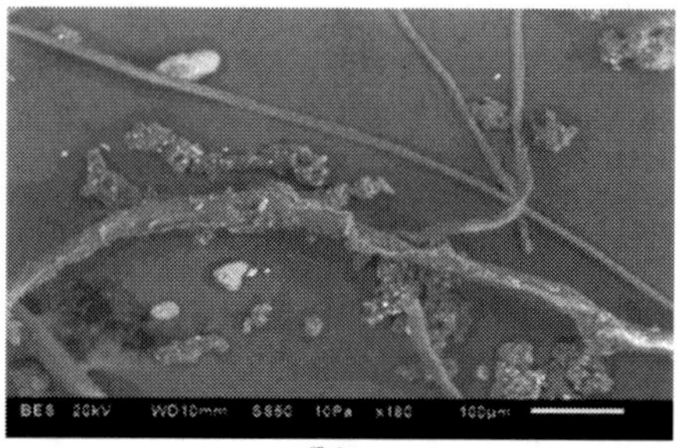

(b)

Figure 4. SEM images of recycled-tire fibers at: (a) 1000x, (b) 180x.

4. Conclusions

The flexural and compressive strength were modified with the addition of recycled fibers, these properties have a concentration dependence of reinforcements utilized. Mechanical properties of PC are improved when the fiber concentration is enough to decrease the negative effect of a poor elastomer-matrix adherence but with a percentage of reinforcement less than 1 vol%. A more ductile material is obtained at expense of the flexural and compressive strength.

Acknowledgments

To Autonomous University of the State of Mexico grant #UAEM 3054/2011SF, and CONACYT for the scholarship support for the achievement of this research. To Faculty of Engineering of the University of Porto (FEUP) for their help in the elaboration and development of this study.

References

[1] Nóvoa J R O P, Ribeiro C S M and Ferreira J M A 2004 *Materials Science Forum* **455-456** 805-809
[2] Bobadilla-Sánchez E A, Martínez-Barrera G, Brostow W and Datashvili T 2009 *eXPRESS Polymer Letters* **3** 615-620.
[3] Martínez-Barrera G, Martínez-Hernández A L, Velasco-Santos C and Brostow W 2009 *e-Polymers* **103** 1-14.
[4] Martínez-Barrera G, Vigueras-Santiago E, Gencel O, Hagg Lobland H E 2011 *J. Mater. Ed.* **33** 37-52.
[5] Martínez-Barrera G, Giraldo L F, López B L and Brostow W 2008 *Polym. Compos.* **29** 1244-1251.
[6] Wang Y, Wu H C and Li V C 2000 *J. Mater. in Civil Eng.* **12** 314-319
[7] Mis-Fernandez R, Azamar-Barrios J A and Rios-Soberanis C R 2008 *J. Appl. Res. Tech.* **6** 95-105
[8] Figueroa-Infante A S, Sánchez-Castillo A and Reyes-Lizcano F 2007 *Épsilon* **9** 41-55

Third Congress on Materials Science and Engineering (CNCIM-Mexico 2012) IOP Publishing
IOP Conf. Series: Materials Science and Engineering **45** (2013) 012027 doi:10.1088/1757-899X/45/1/012027

Effect of the Loofah-fibers on the mechanical properties of Polymer Concrete

M Martínez-López [1], G Martínez-Barrera [2]* and E Martínez-Cruz [1]

[1] Posgrado en Ciencia de Materiales, Facultad de Química, Universidad Autónoma del Estado de México, Toluca México.

[2] Laboratorio de Investigación y Desarrollo de Materiales Avanzados (LIDMA), Facultad de Química, Universidad Autónoma del Estado de México, Km.12 de la carretera Toluca-Atlacomulco, San Cayetano 50200, México. E-mail: gonzomartinez02@yahoo.com.mx

Abstract: Polymer Concrete was reinforced with Loofah fibers in order to improve the compressive and flexural strength. Polymer concrete specimens were prepared with 70% of silicious sand, 30% of polyester resin and various fiber concentrations (0.3, 0.6 and 0.9 vol%). The results show decrease in the values of mechanical properties including flexural strength, compressive strength and compression modulus of elasticity.

1. Introduction

In early civilizations natural polymers combined with inorganic materials were used to produce high strength construction materials. Natural polymers as albumen, blood and rice were used in ancient mortars; the 'asphalt' a natural polymer was used in Babylonia in the mortars and clay brick walls; while in Harappa bituminous mortars were used, or the most famous "glutinous rice paste lime" used to build the Great Wall of China [1- 3].

Concrete is the most widely used structural material in the world, due to its easy preparation and low cost. Some disadvantages on concrete are: a) the pores which becoming entrance points for water, water vapor, gases and chemical substances that could damage to concrete; b) the roughness of concrete surface deteriorates rapidly due to its high abrasion; c) poor resistance to aggressive substances and salty water; d) poor resistance to heat. One alternative for such problems is the incorporation of polymer materials into concrete. They have good binding with mineral aggregates. Combining these with organic compounds, promotes the development of polymer-based materials with wide properties. Moreover, with synthetic polymers is possible to produce cheaper and more flexible materials [4].

Polymer concrete (PC) is a particulate composite where thermoset resins binds inorganic aggregates instead of the water and cement binder typically used in Portland cement concrete (PCC). Polymer concrete consists of a polymeric matrix and dispersed particles of strengthening phases. The polymers constitute the continuous phase hence the composite behavior is largely determined by the properties of the polymer, which are dependent on time, structure and temperature [5]. These composites have some advantages compared to ordinary cement concrete such as: fast hardening, high mechanical strength and high chemical resistance. The features of polymeric concrete depend on the polymer properties, type of filler and aggregates, and the component concentrations. Its applications are continuously diversifying including the building cladding dating from 1958. In recent years the applications include: cultured marble for counter tops, as repair material, overlays for bridge and floors, sport arenas and stadiums, laboratories, hospitals, factories; also precast polymer concrete is used for drains, underground boxes, manholes, acid tanks and cells, tunnel lining, shells, floor tiles, architectural moldings and machine tools and bases.

IOP Conf. Series: Materials Science and Engineering **45** (2013) 012027 doi:10.1088/1757-899X/45/1/012027

Polymer binder can be a thermoplastic, but more frequently a thermosetting polymer. The most frequent polymers are: methyl methacrylate, polyester prepolymer-styrene, epoxyde prepolymer-hardener and furfuryl alcohol.

The polymer concrete advantages versus conventional hydraulic concrete include: Lightweight (1/10 to 1/3 of the concrete weight); high compressive, flexural and tensile strengths (3 to 5 times higher than conventional concrete); reduced installation costs by its easy handle; stable under freeze/thaw conditions; higher impact resistant; low water absorption (less than 1%); resistant to alkalines, acids, weathering and other forms of deterioration; outperforms conventional materials for longer service and has lower life cycle costs; nonflammable and nonconductive.

In order to improve the mechanical strength and the control on the cracks, fibers are added to the polymer concrete. Historically the natural fibers were used for reinforce building materials, but with the arrival of the plastic era, the synthetic fibers shift to natural fibers. Nowdays many scientific and technological efforts are developing for to use again natural fibers as reinforcements. Many natural reinforcing materials can be obtained at low cost by using manpower and technical know-how. Recent studies show that the addition of natural fibers on hydraulic concrete does not significantly influence on the compressive strength but the tensile strength, flexural strength and toughness are substantially improved. The most important contribution of the fibers can be evaluated by determining the fracture toughness of the composite [6-8]. Substantial experience and broader knowledge of the optimal compositions, properties and stress-strain relationships of the fiber-reinforced polymer concrete are necessary with respect to design, production and quality control. However, the data on polyester and epoxy polymer concrete are rather limited, and there is an increasing interest in the deformation characteristics under working conditions. Some studies include chopped strand glass-fibers for to improve the strength and the crack propagations [9].

Natural fibers (coconut, banana, sugar cane bagasse and cellulose) and glass, carbon or boron fibers have been added to polymer concretes. In particular the coconut fibers are excellent reinforcements, and the crack propagation can be controlled by addition of short glass or carbon fibers. A new natural fiber that has attracted attention is the luffa fiber (60% cellulose, 30% hemicellulose and lignin 10%) due to its surface area, low specific gravity, tensile strength (29.4 MPa) and elasticity module (3.4 GPa) [10,11].

In this work the compressive and flexural strength, as well as the modulus of elasticity were measured on polymer concrete reinforced with loofah fibers. The influence of fiber reinforcement was investigated.

2. Experimental

2.1. Specimen preparation

The polymer concrete specimens were elaborated by mixing silica sand SP55-Sibelco (as mineral aggregate with average diameter (d_{50}) of 245 μm and 2.65 g/cm^3 of density) and a commercial unsaturated polyester resin (orthophtalic) in a 70/30 volume ratio. The resin is a viscous liquid with a styrene monomer content of 30% (Aropol™ FS 3992, Matexplás, Porto, Portugal). The proportions of the methyl ethyl ketone peroxide (MEKP) initiator added to the polymer for initiating the free radical polymerization process was 1 mL/by 100 g of the resin weight.

Four different lots of polymer concrete were prepared according to the loofah-fiber (average diameter of 245 μm) concentration, namely 0, 0.3, 0.6 and 0.9 vol%. Each lot has six specimens, meaning 24 polymer concrete specimens for each test. After mixing, the polymer concrete specimens (4x4x16 cm according to RILEM CPT/PC-2 standards) were placed in a controlled temperature room at 23.0 ± 3.0°C for 24 hours.

Mechanical Tests

The compressive and flexural tests of the polymer concrete specimens were carried out in an Instron Universal Testing Machine, with a load cell of 100 kN, according to UNE 83821:1992 test standard and RILEM standards flexural analyses (three-point bending tests).

3. Results and Discussion

3.1. Compressive strength

In Figure 1a we show the compressive strength σ_c values for polymer concretes elaborated with different loofah fiber concentrations. The values diminish when increasing the fiber concentration. The lowest value, namely 19 MPa is 30% less than that for polymer concrete without fibers. Such lowest value was obtained by polymer concrete with 0.9 vol% of fiber.

Some considerations are involved in the diminution of the compressive strength of the polymer concretes: a) When the polymer concrete is subject to compression, a major number of fibers contribute to increase the number of cracks and consequently the numbers of failures, thus lower compressive strength is obtained; b) High porosity and large liquid absorption capacity (13.6 g/g) of the loofha fibers contribute to diminish the compressive strength; c) Unlike other fibers, such as steel or polypropylene, the loofah fibers do not have uniform characteristics as diameter or length, due to its own nature. This anisotropy in loofah-fibers affects significantly when applied the load.

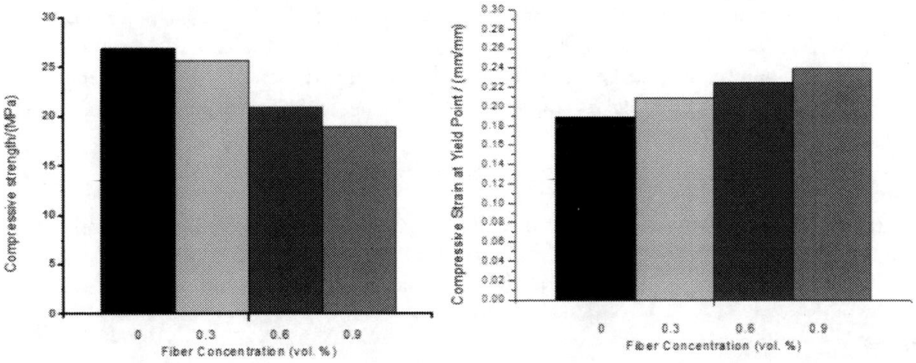

a) b)

Figure 1. Polymer concrete with Loofah fibers: a) Compressive strength, b) Compressive strain at Yield Point.

3.2. Compressive strain at yield point

The compressive strain at yield point values are shown in Figure 1b. The values are increasing when the fiber-loofah concentration increases. The highest value, namely 0.24 mm/mm is 21% higher than those for non-fiber polymer concrete. The compressive strain values have an opposite behavior than those for the compressive strength (as one increases the other decreases).

3.3. Compression Modulus of Elasticity

The values of the compression modulus of elasticity E_c are decreasing when the loofah-fiber concentration increases (Figure 2a). Similar behavior was observed for the compressive strength values. The lowest value (1.89 GPa) is 33% minor than those obtained by non-fiber polymer concrete.

IOP Conf. Series: Materials Science and Engineering **45** (2013) 012027 doi:10.1088/1757-899X/45/1/012027

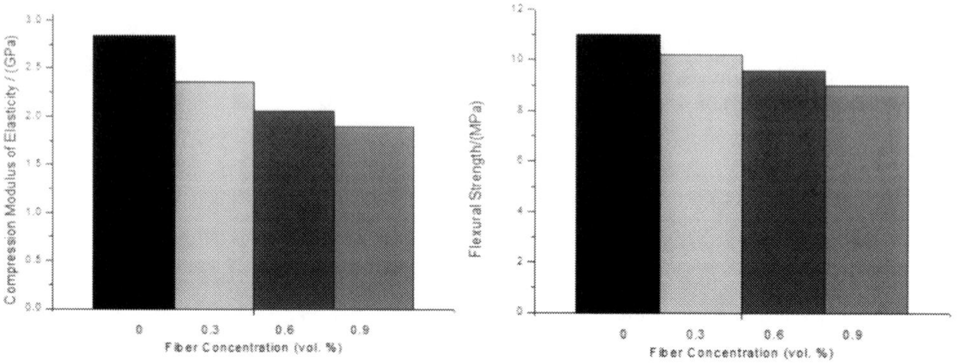

a) b)

Figure 2. Polymer concrete with Loofah fibers: a) Compression modulus of Elasticity, b) Flexural strength.

3.4. Flexural Strength

The flexural strength values of the polymer concrete with looffa-fibers are shown in Figure 2b. The values decrease when loofah-fiber concentrations are increasing. The lowest value, namely 9 MPa, is 18% minor than those obtained by non-fiber polymer concrete.

4. Conclusions

By adding loofah fibers to polymeric concrete it was possible to diminish the compressive strength (42%); flexural strength (23%) and Young's modulus (48%). Nevertheless, the compressive strain was increasing (26%). Such results can be attributed to the characteristics of the loofah fiber, high porosity and water absorption capacity. The fiber pores act as "flaws" in the polymer concrete, thus adhesion between the resin and sand decrease.

Acknowledgments

To Autonomous University of the State of Mexico grant #UAEM 3054/2011SF and CONACYT for the scholarship support for the achievement of this research. To Faculty of Engineering of the University of Porto (FEUP) for their help in the elaboration and development of this study.

References

[1] Wahby W S 2003 *ACI International* **214-2** 13-14.
[2] You-Yun W 1994 *Polymers in Concrete* **12** 55-84.
[3] You-Yun W 1981 *3rd International Congress on Polymers in Concrete* **3** 46-55.
[4] Nielsen L 1974 *Mechanical Properties of Polymers and Composites* **2** 74-86.
[5] Nóvoa J R O P, Ribeiro C S M and Ferreira J M A 2004 *Materials Science Forum* **455-456** 805-809.
[6] Pakotiprapha B 1983 *Behaviour of Bamboo Fiber-Cement Paste Composites* **3** 235-248.
[7] Uzomaka O J 1996 *Characteristics of Akwara as Reinforcing Fiber* **28** 162-167
[8] Campbell M D 1990 *Wood Fiber-Reinforced Cement Composites* **15** 1962-1970.
[9] Vipulanandan C and Dharmarajan N 1998 *Polym Eng Sci* **18** 1182–1191.
[10] Mazali I O 2005 *Morphosynthesis: high fidelity inorganic replica of the fibrous network of loofa sponge (Luffa cylindrica)* **77** 25-31.
[11] Rowell R M 2002 *Characterization and factors effecting fibre properties, In Natural polymers and agrofibres based composites* 115-134.

IOP Conf. Series: Materials Science and Engineering **45** (2013) 012028 doi:10.1088/1757-899X/45/1/012028

Threefold representation of M-N-O systems as a guide to predict mechanosynthesis of M-N nanocrystals

H. Rojas-Chávez[1, 5a], F. Reyes-Carmona[2], L. Huerta[3] and D. Jaramillo-Vigueras[4]

[1] Instituto Tecnológico de Tláhuac – II, Camino Real 625, Col. Jardines del Llano, San Juan Ixtayopan. Del. Tláhuac, México, D.F. 13508
[2] Facultad de Química – UNAM, Circuito de la Investigación Científica s/n, C.U. Del. Coyoacán, México, D.F. 04510
[3] Instituto de Investigación en Materiales – UNAM, C.U. Del. Coyoacán, México, D.F. 04510
[4] Centro de Investigación e Innovación Tecnológica – IPN, Cerrada de CECATI s/n, Col. Santa Catarina, Del. Azcapotzalco, México, D.F. 02250
[5] Centro de Investigación en Ciencia Aplicada y Tecnología Avanzada – IPN, Legaria 694, Col. Irrigación, Del. Miguel Hidalgo, México, D.F. 11500

[a] rojas_hugo@ittlahuac2.edu.mx

Abstract. A method is proposed to forecast how micrometric powders are converted on intermetallic nanocrystals by high-energy milling. Mixtures of MO-N or M-N as precursors were processed to obtain PbTe or PbSe. This transformation takes place by oxidizing highly reactive chalcogens such as Se or Te and reoxidation of PbO to form Pb complex chalcoxides. In the latter phases Te or Se share electrons to act as Te^{IV} or Se^{IV}, $PbTeO_3$ or $PbSeO_3$. Once that state is completed, the reduction state manifests itself as an amorphous precipitation of phases. The final stage is traced as high purity nanocrystals.

1. Introduction

In view of a wide world of applications of intermetallic chalcogenides, a great effort is brought to researchers to explore and develop new process routes to produce efficiently high purity nanocrystals. Results have been published recently on several process routes, among those are: thermal decomposition of precursors, selective vapor deposition, organometallic chemical treatment of reactive species, controlled fusion and solidification of pure metals to render nanocrystalline phases (by Czochralki method) and high-energy milling (HEM) [1-2].

Undoubtedly, among those processes one must recognize that the last one is simpler than any other; and the high quality and purity of intermetallic compounds can be reached, provided physical and chemical parameters are controlled [1, 3].

It must be established clearly that by HEM one can produce alloying and mechanosynthesis of intermetallic phases. This means that by milling pure metals like Pb and Te or Se, or alternatively by milling mixtures of PbO and either Te or Se one can produce nanopowders of the type PbTe or PbSe, respectively. Under any of those circumstances, there are two basics mechanisms by which the transformation is induced, namely: themo-mechanical and physic-chemical driving forces.

Therefore, to avoid presence of impurities a set of ZrO_2-spheres and a nylamid vial would be very favorable system components to be considered in the HEM-process. That is due to high impact resistance, wear resistance and also owning to their inert chemical nature [3-4].

A critical chemical factor to be considered in the synthesis of PbO + Te or PbO + Se, as expected; it is to eliminate the effect of the final oxygen potential. An implicit reason considered is that an oxide phase should dictate the maximum surface oxygen potential [1].

If the mechanosynthesis via the HEM is not underestimated as a process then, a sequence of effects can be described as: physical and chemical joined powders that constitute agglomerates as well as an incipient presence of chalcogenides as intermetallic compounds [1]. These events take place since very early in the grinding process; and so they indicate us that friction, fracture and low energy collisions give way to both Van der Waals junction forces and local transformation of intermetallic phases [1, 3].

Lead intermetallic chalcogenides such as PbTe or PbSe as well as pure Te or Se are highly reactive species. These acts as oxygen scavengers and their reactivity are favored by HEM [5].

Several researchers have evaluated a wide range of stoichiometric chemical complex compounds that can be framed within ternary ratios in the Pb–Te–O or Pb–Se–O systems. Among them, one can list the following: $PbTeO_3$, $PbTeO_4$, $PbTe_5O_{11}$, $Pb_2Te_3O_8$, Pb_2TeO_4, Pb_3TeO_5 and Pb_5TeO_7. And, for complex selenium oxides almost equivalent phases including $PbSeO_3$ have also been reported [5-6]. Additionally, simpler species related to complex oxides are: 'Pb_2O' (it was traced as an amorphous non-stoichiometric compound), PbO, PbO_2, TeO_2, SeO_2 and TeO_3 (PbO_2 and TeO_3 were scarcely found; however their index of detection was unambiguously traced). Other oxides traced experimentally are those that do not necessarily obey stoichiometric ratios between Pb, Te and O or Pb, Se and O; instead amorphous or even vitreous phases can be traced.

Another set of experimental and theoretical data has been reported on lead telluride reactivity measured in respect of different oxygen partial pressures [5-6]. Thus, the following set of reactions was postulated

$$PbTe + 2\,O_2(g) = PbTeO_4 \qquad (1)$$

$$PbTe + \tfrac{3}{2}O_2(g) = PbTeO_3 \qquad (2)$$

$$PbTe + O_2(g) = PbTeO_2 \qquad (3)$$

2. Development of the proposal

The first hypothesis under which the experimental campaign was designed was that where a mixture of a chemically stable oxide plus an avid element for oxygen, under the influence of HEM, one might induce an overall reducing effect. Thus, mechanical energy dissipated as heat, fracture in addition of plastic deformation could become driving forces to start transformation of precursors onto intermetallic phases [1, 3-4]. Henceforth, an ideal way to represent that proposal would be represented by:

$$MO(s) + N(s) \xrightarrow{\Delta E_m} MN(s) + \tfrac{1}{2}\,O_2(g) \qquad (4)$$

where: M represents Pb and N can be Te or Se.
ΔE_m stands for energy supplied by HEM.

Since micropowders of MO and N are subjected to random collisions rates, only the most efficient impacts are expected to act as a spark to initiate chemical transformations. This specific reaction condition is reached in the milling system when both temperature and gaseous chalcogens phases are in close contact with ultrafine or nanosized PbO powders. Thus, provided there are no other oxides and solely PbO is reduced, then as thermochemically expected from any intermetallic formation, energy would be continuously released, therefore a thermally self-sustained process would be expected [1].

IOP Conf. Series: Materials Science and Engineering **45** (2013) 012028 doi:10.1088/1757-899X/45/1/012028

Under these process conditions, a chemical starting point would be considered as the initial transient condition that will give way to a steady state transformation process.

However, such proposal is partially valid, since evolution of chemical conditions traced during milling indicate us that in addition of PbO other single and complex oxides can also be present [1-4]. Due to its importance, this finding will be closely approached in a further section.

Other research works along these lines, have also proposed that processing pure metals to obtain lead chalcogenide can be represented by the following reaction [5, 7]:

$$M(s) + N(s) \xrightarrow{\Delta E_m} MN(s) \quad (5)$$

To this point, it is important to cite that out of the MN(s) phases obtained by the HEM, it is expected to reach very high purity nanocrystals. Therefore, a massive transformation from pure metals, as given by reaction (5), does not necessarily satisfy that stringency mentioned previously. It is acknowledged that very small oxygen partial pressures as those reached under experimental atmosphere conditions, would decidedly affect precursors as well as products considered in reactions (1-5).

In view of these findings, a modified reaction scheme was proposed to take into account an intermediate stage. This proposal is represented by the following scheme, as shown in figure 1.

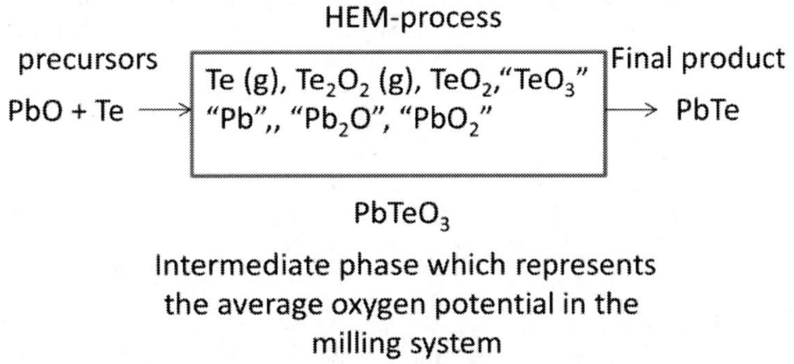

Figure 1. Scheme of the HEM process considered as a batch reactor.

3. Experimental and theoretical results

By comparing mechanical metallurgical versus physicochemical responses in the HEM-process, it was found that the former factor was quite modest as compared to the later one. In view of that, an acute analysis was centered on chemical transitions induced by thermally activated process.

The first and decisive point to be considered about the physicochemistry of transformations from micropowders as precursors to nanocrystalline phases of high purity is that: (i) powders were charged under atmosphere conditions (this statement means that an initial oxygen potential given by PbO plus Te was already present in the system) and (ii) it was not attempted to remove humidity from ZrO_2-spheres neither from internal walls of the vial [3-4]. Therefore, reasons to perform experiments through these conditions were that an oxide (PbO) was as one of the precursors.

During milling through which precursors were subjected, to achieve the mechanosynthesis of either PbTe or PbSe via the HEM, gaseous phases enriched in either Te or Se were identified. These appeared on surfaces of the milling media (ZrO_2-spheres). Appearances of these deposits were either bright metallic or grayish, thus indicating a condition of either metal or oxide vapors [8], see figures 2 and 3. By processing PbO plus Te, it is expected to relate those phases with Te_2 (g) or Te_2O_2 (g) while the twofold system PbO plus Se they correspond to Se_6 (g) and SeO (g) [9-10]. As a consequence, the partial pressure of chalcogen gaseous phase should override gradually the oxygen partial pressure in the vial [1].

Additionally, regarding early tracing of intermetallic phases, it is a clear indication as to how a chemical transition from PbO and Te is promoted by the fractured surface in contact with chalcogens gaseous phases [3].

Thermochemical stabilities of these phases were confirmed by identifying them with data reported in the literature [9-10]. Temperature ranges through which theoretical data was disclosed were those traced by an infrared thermometer; the overall highest temperature registered was around 550 K.

Based on other reports on thermal responses during milling an educated guess would lead us to propose that powders could reach thermally as twice as that value [1, 11].

Among the complex oxide phases plentifully identified in the Pb–Te–O system was the stoichiometric $PbTeO_3$ phase. In the Pb–Se–O system, as shown in figure 4, the equivalent phase was $PbSeO_3$, other complex phases in the PbO–TeO_2 subsystem have a quasi-similar composition; however, they were amorphous and did not have well defined Pb:Te:O composition ratios [1, 3].

By using either conventional analytical techniques such as XRD or electron microscopy techniques as SEM, TEM or HRTEM and even XPS as well, other phases like PbO_2, TeO_2 and SeO_2 were identified [1, 3, 8, 12], see figure 4. Additionally, under high (pulse) energy discharges by the TEM-electron gun, non-crystalline as well as non-stoichiometric phases in terms of Pb:Te:O or Pb:Se:O ratios were induced. These species, however, showed ratios closely related to $PbTeO_3$ or $PbSeO_3$, respectively.

Just before any discussion is derived out of these findings and also previously to make any proposal related to a chemical mechanism; we believe, it is very important to establish that chemical analysis referred to oxides, others than PbO, are traced on surfaces of powders. It means that any composition effects between precursors, gases species and powders and between gases (oxygen and gaseous chalcogens) solely represent local and superficial powder chemical changes [8].

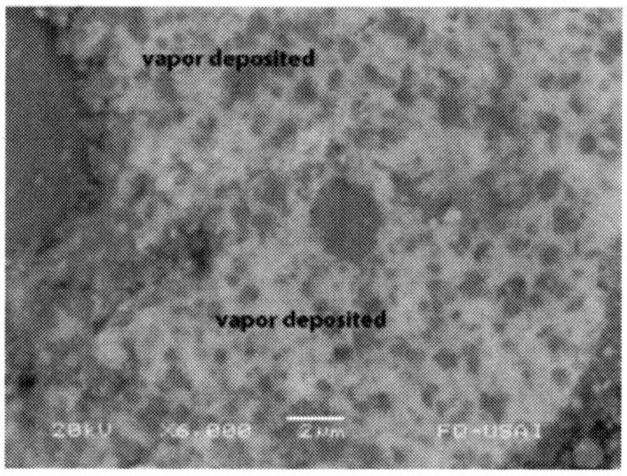

Figure 2. Gray oxide deposited on ZrO_2-spheres as vapor during milling of PbTe nanoparticles.

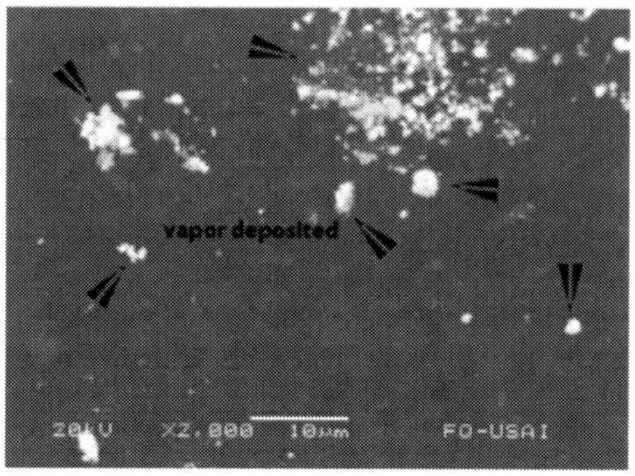

Figure 3. Bright oxide deposited on ZrO_2-spheres as vapor during milling of PbSe nanoparticles.

Through these postulates, one can infer that if mixtures of precursors such as PbO plus both Te or Se then plastic deformation and fracture as well as chalcogen vapor phases may be found during HEM [1, 8]. Therefore, if the thermal potential becomes a strong driving force to induce chemical transition then fresh fractured surfaces, in close contact with gaseous chalcogens phases, will become local efficient mass exchangers.

4. Development of the proposal or model

To frame all possible stable single and complex phases, as earlier was proposed to account for the Pb–Se–O system [1], a ternary system was disclosed, see figure 5. There, pure substances were located on their apexes. Each side of the triangle includes three binary sides namely: Te–O, Pb–O and Te–Pb. The Te–O side includes phases like ½ Te_2O_2, TeO_2 and TeO_3. The Pb–O axis includes PbO and PbO_2; and the remaining Pb–Te axis, solely includes the PbTe. However, the most important subsystem is that relating two oxides, PbO and TeO_2. Thus, by using an equivalent analogy, it becomes evident that an analysis based on compositions of precursors as well as on data shown in the figures 1–4 would not be enough to propose a chemical model. However, by taking, for example all data previously disclosed as single and complex oxides referred to the Pb–Te–O system then a PbO–TeO_2 subsystem can be used to frame most of the complex oxide phases traced in this work, see figure 5. Through this approach a very clear sequence of changes that might evolve during milling can be located as stoichiometric phases, see figure 5. Additionally, other secondary subsystems like those where the dimer gaseous phase, ½ (Te_2O_2), is related to PbO_2 or where the intermetallic, PbTe, is connected to pure oxygen or the richest oxide, TeO_3, related to pure Pb, all of them can be adequately represented in the Pb–Te–O ternary system, as shown in figure 5.

If a closer analysis is carried out, on the ternary representation of this system, it is evident that all of the most common species related between those extreme chemical couples (a–c), is the $PbTeO_3$ phase.

As earlier described, if phase transformations are evaluated as cumulative collision as well as statistical cumulative chemical events, one can postulate that the $PbTeO_3$ phase can be the average chemical effect which in turn will determine the average oxygen potential.

It is worth to emphasize that through most of the milling process this phase was that which satisfies the stoichiometric ratios of 1:1:3, respectively. This finding clearly reveals that tellurium in this phase is acting as a tetravalent atom.

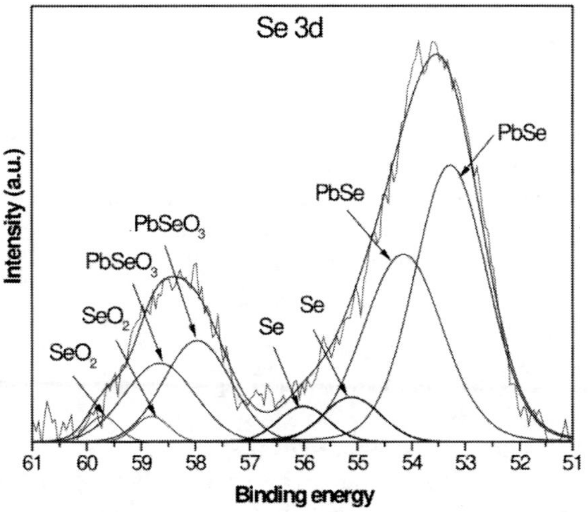

Figure 4. XPS spectrum of Se 3d core level [8].

To complete a description of the physicochemical role played by the $PbTeO_3$, a dynamic equilibrium is proposed where the threefold phase set: PbO–PbTe–$PbTeO_3$ could coexist during milling. It is worth to point out that an equivalent approach was tested and proved for the triple set of components for the PbO–PbSe–$PbSeO_3$ system [1], see figure 4. The only difference found between these systems is that the gaseous phase in the Pb–Te–O system involves ½ Te_2O_2 (g) instead SeO (g) and Te_2 (g) instead Se_6 (g), see figure 4.

This assemblage of values constitutes a thermodynamical arrangement that here represents the quasi-steady state transformation. Hence, most of the milling process will be regulated by either the PbO–PbTe–$PbTeO_3$ or the PbTe–$PbTeO_3$–½Te_2O_2, as dictated by the Gibbs phase rule [8]. That inner

IOP Conf. Series: Materials Science and Engineering **45** (2013) 012028 doi:10.1088/1757-899X/45/1/012028

triangle shown in figure 6 represents those dynamic equilibrium condition developed during most of milling time.

So, if the maximum oxidation potential is that related to the presence of $PbTeO_3$, the reduction condition will have to be considered as that where the corresponding oxygen potential is overcame by gaseous chalcogens partial pressures.

Thus, to account for a chemical transition from high to low oxygen potential a free oxygen effect may be related through a reduction along the PbO–TeO_2 extremes, where the 2 PbO/TeO_2 ratio is obeyed. See point (5) in figure 5.

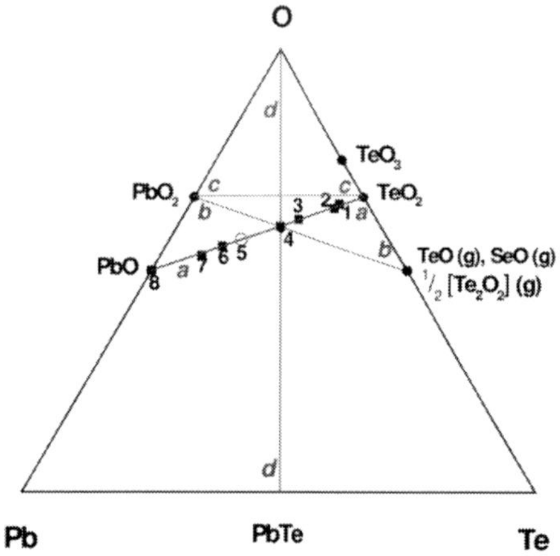

Figure 5. Ternary representation of powder products at different milling times.

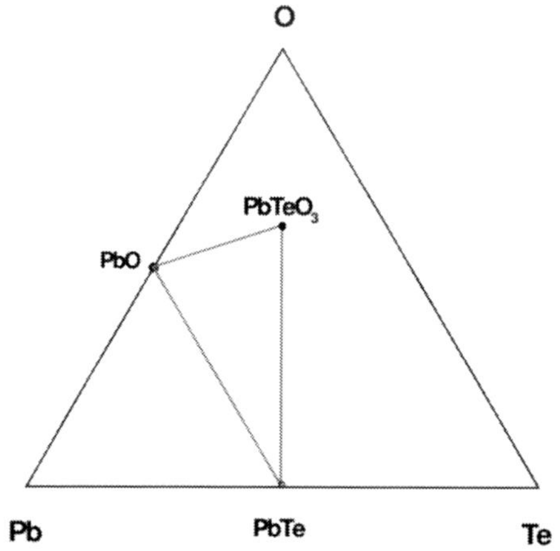

Figure 6. Ternary representation of the dynamic equilibrium condition developed during milling.

The last step to which nanopowders are subjected to would be those where highly crystalline phases are obtained. It is quite feasible that this final transition stage can be attributed to the total removal of the oxygen potential effect. Therefore, if partial pressures of oxygen are diminished to their lowest extent, it means that partial pressures given by the chalcogens gaseous phases would be definitely larger than oxygen atmospheric pressure.

5. Final remarks

By HEM microsize precursors like PbO in addition of either Te or Se can mechanosynthesized to render intermetallic compounds like PbTe or PbSe, respectively.

Through this process, PbO is incipiently reduced by Te, and as expected Te is slightly oxidized during early stages of milling. Once, energy conditions due to effective collisions between powder and milling media rise the temperature of the milling system, gaseous chalcogens phases like Te_2 (g), Te_2O_2 (g) and Se_6 (g) as well as SeO (g) are formed in the Pb–Te–O and Pb–Se–O systems, respectively.

However, the effect of the reduction of PbO plus those oxygen contributions of physisorbed or chemisorbed oxygen or humidity as well as that oxygen trapped when powders were charged in the vial, all of them locally promote high oxygen potentials on surface of powders.

The average oxygen potential developed on surface of powders is that represented by the $PbTeO_3$ phase as shown on the ternary diagram in figure 5 (point 4).

Once this oxygen potential is surmounted by either Te_2O_2 (g) or SeO (g) then Te_2O_2 (g)/Te_2 (g) or SeO (g)/Se_6 (g) ratios become quite effective on reducing the $PbTeO_3$ and $PbSeO_3$, respectively.

It is important to point out that the reoxidation phenomenon as an average value represented by the $PbTeO_3$ involves higher valence (4+) oxides like PbO_2 or TeO_2 or even TeO_3 (6+). Additionally, lower oxygen potentials than that given by the $PbTeO_3$ would be phases involved like Te_2O_2 (g) or SeO (g). As a direct consequence of these findings chemically the $PbTeO_3$ can act as a pivot point with changes of Te-valences induces metastable non-stoichiometric phases on transformations.

Additionally, if oxygen content is not fully removed from nanoparticles then crystals would not be faceted, instead these would manifest themselves as quasi-spherical shaped particles.

Thus, highly crystalline nanoparticles will be traced only when even traces of oxygen are totally removed.

A final remark to be mentioned is that related to the kinetics of the mechanosynthesis of these chalcogenides. It is that referred to the presence of Te_2O_2 (g) against Se_6 (g) or SeO (g). Lead tellurides due to Te_2O_2 (g) induces the PbTe rate of transformation in almost one half time that to transform PbSe.

Acknowledgements
One of the authors (H. Rojas-Chávez) would like to thank CONACyT for the *financial support extension*. The authors are grateful to G. Gónzalez-Mancera (FQ-UNAM) and C. Flores-Morales (IIM-UNAM) for their technical support.

References

[1] Hugo R, Fidel R and David J 2011 *Mater. Res. Bull.* **46** 1560
[2] Jalal M and Kelvin G 2008 *J. Crystal Growth* **310** 4217
[3] Hugo R, Fidel R and David J 2010 *Rev.Metal. Madrid* **46 (6)** 548
[4] Hugo R, Sebastián D, David J and Gabriel P 2009 *J. Alloys Compd.* **483** 275
[5] J. Galat, J. Haber, J. Nowotny, et al. 1975 *Oxidation of Metals* **9** (**6**) 497
[6] Lidong C, Takashi G and Toshio H 1997 *IEEE* 16th Inter. Conf. on Thermoelectrics 251
[7] Girard SN, Zhou X, Shoemaker D, et al. 2011 *J. Am. Chem. Soc.* **133** (**41**) 16588
[8] Hugo R, Fidel R and David J 2012 In press
[9] I. Barin 1995 ***Thermochemical data of pure substances*** **3rd Ed** (New York: Weinheim, VCH)
[10] F.J. Mompean, J. Perrone and M. Illemassène 2005 ***Chemical Thermodynamics of Selenium*** **Vol. 7** (The Netherlands: ELSEVIER B.V.)
[11] Park Ch, Won M, Oh Y, et al. 2005 *Applied Surface Science* **252** 1988
[12] Hugo R, Fidel R, Marcela A, et al. 2012 *J. Napart. Res.* **14 (6)** 897

IOP Conf. Series: Materials Science and Engineering **45** (2013) 012029 doi:10.1088/1757-899X/45/1/012029

Theoretical study of the Diels-Alder reaction between *o*-benzoquinone and norbornadiene

Ramiro F Quijano-Quiñones[1], M Quesadas-Rojas[1], Gabriel Cuevas[2] and Gonzalo J Mena-Rejón[1]

[1]Laboratorio de Química Farmacéutica, Facultad de Química, Universidad Autónoma de Yucatán, calle 41 No. 421 x 26 y 28 col. Industrial C.P. 97150. Mérida, Yucatán, México
[2]Instituto de Química, Universidad Nacional Autónoma de México, Circuito exterior, Ciudad Universitaria, C.P. 04510 México D.F., México

E-mail: ramiro.quijano@uady.mx

Abstract. The reaction between norbornadiene and *o*-benzoquinone is an important step in polyalicyclic rigid structures synthesis. It has been considered that this reaction is an example of Diels-Alder (DA) and hetero-Diels-Alder (HDA) cycloadditions with *o*-benzoquinone acting as diene (forming C-C bonds) and heterodiene (forming O-C bonds). We have performed a Density Functional Theory study of this reaction, employing B3LYP, mPW1PW91, and B1B95 functionals and 6-31G(d,p) and 6-31+G(d,p) Gaussian type basis sets. The results indicate that Diels-Alder is a feasible mechanism for both reactions, but should not be the main route to the formation of products with C-C bonds.

1. Introduction

The Diels-Alder (DA) and Hetero-Diels-Alder (HDA) reactions are [4+2] π electrons cycloadditions of great importance for organic chemistry and represent one of the most powerful methods in synthetic chemistry. In particular, the Diels-Alder reaction of *o*-benzoquinone as a diene and norbornadiene as a dienophile has been used to produce rigid polyalicyclic structures [1-3] with the ability to act as supramolecular hosts by acting as molecular tweezers, clamps, and clips for synthetic receptors [4] and have been employed in host-guest studies [5-6].

Margetić *et al.* [7] have studied the reaction of *o*-benzoquinone (R=H, Figure 1) and *o*-chloranil (R=Cl, Figure 1) with norbornadiene. Experimentally, they have found that *o*-chloranil gave six adducts 1-6. In contrast, the similar reaction with *o*-benzoquinone only gives four adducts 1-4 with the adduct 1 being the dominant product, while adducts 2 and 3 were found in smaller amounts, and adduct 4 was detected by spectroscopy, but not isolated. In order to explain the experimentally observed stereospecificities, the author performed a theoretical study of these Diels-Alder reactions using *ab initio* methods at DFT level using a Gaussian type basis set 6-31G(d) and the B3LYP hybrid functional for the exchange-correlation potential. Given their failure to locate concerted transition structure for the HDA cycloaddition of *o*-benzoquinone (adducts 5 and 6), they proposed a second, nonconcerted biradical mechanism leading to formation of the products 5 and 6. However, the performance of the theoretical method was not evaluated, which is important to establish the reliability of theoretical results, since there are evidence that B3LYP [8] method usually underestimates barrier heights [9, 10] and there are several hybrid functionals developed to increase B3LYP performance

incorporating kinetic energy of electrons and/or changing the amount of exact HF exchange in the exchange-correlation functional form [11]. This is the case of mPW1PW91 [12] which incorporates 25% of exact HF exchange (in contrast with 20% included in B3LYP), and B1B95 [13] which includes kinetic energy of electrons. Both exchange-correlation potentials have demonstrated a better performance than B3LYP in prediction of energy barriers of some organic reactions [14-16].
In order to get a better understanding of the reaction nature, we have performed a systematic study using different Gaussian basis sets and different approximations to the exchange-correlation functional, to evaluate changes in the transition state structure and the activation energy of this reaction with respect to the method of calculation.

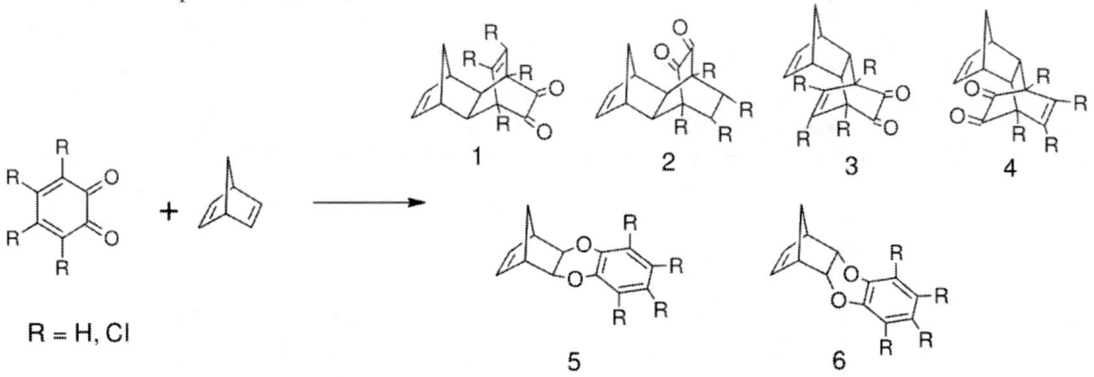

Figure 1. Reaction between *o*-benzoquinone (R=H) and *o*-chloranil (R = Cl) with norbornadiene.

2. Computational Details

All the quantum chemical calculations were performed with Gaussian 09 [17]. Geometrical optimization, including transition-structure searches, was carried out with Gaussian type 6-31G(d,p) and 6-31+G(d,p) basis sets, using Density Functional Theory (DFT) hybrid methods with B3LYP, mPW1PW91, and B1B95 functionals for evaluate the exchange-correlation potential, employing the following convergence criteria: Maximum force on each atom < 0.023 eV/Å, mean root square of the maximum force on each atom < 0.0153 eV/Å, maximum displacement of each atom < 0.0095 Å, and mean root square of maximum displacement of each atom < 0.00064 Å.

Transition-structures search was carried out with Synchronous Transit-Guided Quasi-Newton (STQN) Method, which was developed by Schlegel and co-workers [18], and uses a linear synchronous transit or quadratic synchronous transit approach to get closer to the quadratic region around the transition state and then uses a quasi-Newton or eigenvector-following algorithm to complete the optimization.

Transition states were verified by vibrational analysis, confirming the presence of one imaginary frequency corresponding to a movement in the direction of the reaction coordinate. The values of the activation energies have been calculated on the basis of the total energies of the stationary points.

3. Results and Discussion

All the transition states were located and their degree of asynchronicity (Δd) and activation energy (ΔE) are collected in tables 1 and 2, while transition state structures at B3LYP/6-31+G(d,p) are depicted in figure 2.

Our results (table 1) shows that the transition states TS1, TS5, and TS6 are highly synchronous, with Δd values below 0.05. On the other hand TS2 and TS3 were slightly asynchronous with values up to 0.135, whereas TS4 was highly asynchronous, with Δd up to 0.466. All our attempts to locate a synchronous TS4 were unsuccessful and converged to a non-synchronous TS4.

As we expected from previously reports [7, 14, 16], the size of the barrier depends on the calculation method. Table 2 shows that with mPW1PW91 and B1B95 exchange-correlation functionals, we obtain lower activation barriers than with B3LYP. Taking into account that B3LYP

IOP Conf. Series: Materials Science and Engineering **45** (2013) 012029 doi:10.1088/1757-899X/45/1/012029

underestimates activation barriers [9, 10] we can conclude that mPW1PW91 and B1B95 are poor candidates for the study of this reaction. Furthermore, the effect of adding diffuse functions is increase all activation energies about 1-2 kcal/mol as well as the degree of asynchronicity.

Table 1. Degree of asynchronicity[a] for the conversion of *o*-benzoquinone and norbornadiene to adducts 1-6.

Level of calculation	1	2	3	4	5	6
B1B95/6-31G(d,p)	0.001	0.002	0.039	0.241	0.000	0.003
B1B95/6-31+G(d,p)	0.001	0.011	0.041	0.331	0.001	0.003
B3LYP/6-31G(d,p)	0.001	0.040	0.095	0.316	0.001	0.006
B3LYP/6-31+G(d,p)	0.000	0.135	0.116	0.466	0.000	0.006
mPW1PW91/6-31G(d,p)	0.000	0.001	0.001	0.245	0.001	0.004
mPW1PW91/6-31+G(d,p)	0.002	0.041	0.100	0.298	0.000	0.001

[a] *d*TS/P, defined as the difference between the ratios of the forming bond lengths in the TS and the corresponding bond lengths in the product [19].

Table 2. Calculated activation energies for the conversion of *o*-benzoquinone and norbornadiene to adducts 1-6 (kcal/mol).[a]

Level of calculation	TS1	TS2	TS3	TS4	TS5	TS6
B1B95/6-31G(d,p)	13.8	15.2	15.5	20.3	9.5	12.6
B1B95/6-31+G(d,p)	15.0	16.7	16.8	21.8	11.6	14.3
B3LYP/6-31G(d,p)	18.3	20.0	20.8	24.4	9.6	12.8
B3LYP/6-31+G(d,p)	20.2	22.0	22.9	26.6	12.3	15.1
mPW1PW91/6-31G(d,p)	13.1	15.0	15.3	19.9	8.6	11.7
mPW1PW91/6-31+G(d,p)	14.5	16.5	16.8	21.5	10.7	13.5

[a] Label TS_i refers to the transition state for the formation of adduct i.

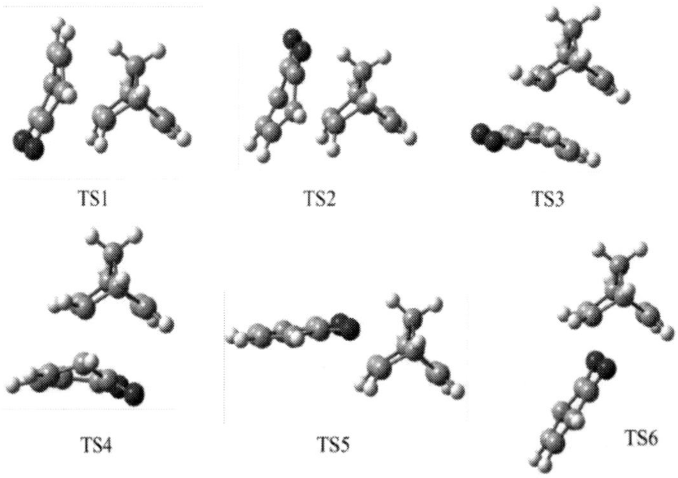

Figure 2. Transition states (B3LYP/6-31+G(d,p)) for the reaction between *o*-benzoquinone and norbornadiene to form adducts 1-6.

All the levels of calculation predict the following order on the activation energies, TS5 < TS6 < TS1 < TS2 < TS3 < TS4, which is not according to the experimental selectivity found for products 1-6 [1, 7]. However, the fact that transition states TS5 and TS6 exist, means that reaction producing adducts 5 and 6 is viable, and could operate by DA mechanism, in contrast with the findings in a previous report [7].

4. Conclusions

We find the transition states TS5 and TS6, nevertheless we are not according with observed experimental selectivity. These results suggest that Diels-Alder is a feasible mechanism for both reactions, but should not be the main route to the formation of products with C-C bonds. There are other alternatives mechanistic paths: ionic or radical mechanism.

Acknowledgments

The financial support was provided by Facultad de Química of Universidad Autónoma de Yucatán, México (PIFI-FOMES 2007).

References

[1] Warrener R N, Johnston M R and Gunter M J 1998 *Synlett.* 593-6.
[2] Johnstom M R, Gunter M J and Warrener R N 2002 *Tetrahedron* **58** 3445-51.
[3] Warrener R N, Johnston M R, Schultz A C, Golic M, Houghton M A and Gunter M J. 1998 *Synlett* 590-2.
[4] Klärner F G, Panitzky J, Preda D and Scott L T 2000 *J. Mol. Model.* **6** 318-27.
[5] Yamaguchi T, Ishii N, Tashiro K and Aida T 2003 *J. Am. Chem. Soc.* **125** 13934-5.
[6] Dudi´c M, Lhoták P, Stibor I, Petˇríˇcková H and Lang K 2004 *New J. Chem.* **28** 85-90.
[7] Margetić D, Johnston M R and Warrener R N 2001 *Molecules* **5** 1417-28.
[8] Becke A D 1988 *Phys. Rev.* A **38** 3098-3100.
Lee C, Yang W and Parr R G 1988 *Phys. Rev.* B **37** 785-789.
Stephens P J, Devlin F J, Chabalowski C F and Frisch M J 1994 *J. Phys. Chem.* **98** 11623-7.
[9] Durant J L 1996 *Chem. Phys. Lett.* **256** 595-602.
[10] Lynch B J, Fast P L, Harris M and Truhlar D G 2000 *J. Phys.Chem.* A **104** 4811-5.
[11] Cohen A J, Mori-Sánchez P, Yang W 2012 *Chem. Rev.* **112** 289-320.
[12] Adamo C and Baron V 1998 *J. Chem. Phys.* **108** 664-675.
[13] Becke A D 1996 *J. Chem. Phys.* **104** 1040-6.
[14] Barquera-Lozada J E and Cuevas G 2009 *J. Org. Chem.***74** 874-83.
[15] Matsuda S P T, Wilson W K and Xiong Q 2006 *Org. Biomol. Chem.* **4** 530-43.
[16] Zao Y, Pu J, Lynch B J and Truhlar D G 2004 *Phys. Chem. Chem. Phys.* **6** 673-6.
[17] Gaussian 09, Revision A.1, Frisch, M. J.; Trucks, G. W.; Schlegel, H. B.; Scuseria, G. E.; Robb, M. A.; Cheeseman, J. R.; Scalmani, G.; Barone, V.; Mennucci, B.; Petersson, G. A.; Nakatsuji, H.; Caricato, M.; Li, X.; Hratchian, H. P.; Izmaylov, A. F.; Bloino, J.; Zheng, G.; Sonnenberg, J. L.; Hada, M.; Ehara, M.; Toyota, K.; Fukuda, R.; Hasegawa, J.; Ishida, M.; Nakajima, T.; Honda, Y.; Kitao, O.; Nakai, H.; Vreven, T.; Montgomery, Jr., J. A.; Peralta, J. E.; Ogliaro, F.; Bearpark, M.; Heyd, J. J.; Brothers, E.; Kudin, K. N.; Staroverov, V. N.; Kobayashi, R.; Normand, J.; Raghavachari, K.; Rendell, A.; Burant, J. C.; Iyengar, S. S.; Tomasi, J.; Cossi, M.; Rega, N.; Millam, N. J.; Klene, M.; Knox, J. E.; Cross, J. B.; Bakken, V.; Adamo, C.; Jaramillo, J.; Gomperts, R.; Stratmann, R. E.; Yazyev, O.; Austin, A. J.; Cammi, R.; Pomelli, C.; Ochterski, J. W.; Martin, R. L.; Morokuma, K.; Zakrzewski, V. G.; Voth, G. A.; Salvador, P.; Dannenberg, J. J.; Dapprich, S.; Daniels, A. D.; Farkas, Ö.; Foresman, J. B.; Ortiz, J. V.; Cioslowski, J.; Fox, D. J. Gaussian, Inc., Wallingford CT, 2009.
[18] Peng C and Schlegel H B 1993 *Isr. J. Chem.* **33** 449-54.
[19] Legnani L, Lunghi C, Albini M F, Nativi C, Richichi B and Toma L 2007 *Eur. J. Org. Chem.* **2007** 3547-54.

IOP Conf. Series: Materials Science and Engineering **45** (2013) 012030 doi:10.1088/1757-899X/45/1/012030

Acoustic transmittance of an aperiodic deterministic multilayer structure

J Madrigal-Melchor, A Enciso-Muñoz and D A Contreras-Solorio

Unidad Académica de Física. Universidad Autónoma de Zacatecas. Calzada Solidaridad esquina con Paseo la Bufa S/N, C.P. 98060. Zacatecas, Zac. México

Email: dacs10@yahoo.com.mx

Abstract. We study theoretically the acoustic transmission for a multilayer structure where the characteristic acoustic impedance follows the values generated by the self-similar sequence called the "1s counting sequence". The transmission spectrum shows clearly self-similarity characteristics.

1. Introduction

The 2011 Nobel Prize in Chemistry was awarded to Shechtman for the discovery of the quasicrystals [1]. After that work, there have emerged in the literature many types of deterministic aperiodic structures, which constitute a very interesting field of research, such as one-dimensional quasiperiodic multilayers. These systems present properties not observed in periodic or completely disordered structures. Merlin *et al* [2] were the first to grow a Fibonacci lattice of GaAs and AlAs and they studied its x ray diffraction and Raman scattering properties. Since then, many other interesting studies of propagation of electrons, electromagnetic waves and acoustic waves have been reported in quasiperiodic multi-layered one dimensional structures with several profiles like those of Fibonacci, Cantor, Rudin–Shapiro, Thue–Morse, and others [3–5]. Of particular interest is the understanding of the propagation and transmission of elastic waves in this type of structures due to the intrinsec theoretical value and also in the understanding of the propagation of seismic waves in multilayer systems and for technological applications [6]. In this work we calculate the transmittance of longitudinal elastic waves for an aperiodic multilayer structure where the characteristic acoustic impedance follows a self-similar numerical sequence named 'the 1s-counting sequence', formed by the number of 1s in the binary representation of the natural numbers [7]. In an article published formerly [8] a calculation was made of the optical transmission for a multilayer structure where the refractive index is modulated by this sequence. We use for the calculations the transfer matrix method.

2. Theoretical model and method of calculation

We consider longitudinal elastic plane waves coming from the left on the x axis in a medium of acoustic impedance Z_0. The wave is incident normal to a structure of N plane layers. After going through the structure, the waves propagate in another medium R with impedance Z_R. Each j layer has width d_j, impedance Z_j and is homogeneous, isotropic and losseless. The characteristic acoustic impedance is given by

$$Z_j = \rho_j c_j \tag{1}$$

where ρ_j and c_j are, respectively, the density and the sound speed for the j material. The wave function is written in each layer as

$$F_j = A_j e^{i(k_j x - \omega t)} + B_j e^{i(-k_j x - \omega t)} \tag{2}$$

F_j represents the traveling pressure perturbation. The amplitudes A_j and B_j represent propagation to the right and to the left, respectively. $j = 0$ is the medium at the left side of the structure with impedance Z_0. In the medium R with $j = N + 1$ we consider only propagation to the right, consequently $B_R = 0$. t is the time, ω the angular frequency and i the imaginary unit. The propagation from medium 0 to medium R through the multilayer structure is described by

$$\begin{bmatrix} A_0 \\ B_0 \end{bmatrix} = \begin{bmatrix} M_{11} & M_{12} \\ M_{21} & M_{22} \end{bmatrix} \begin{bmatrix} A_R \\ 0 \end{bmatrix} \tag{3}$$

The (2 × 2) transfer matrix is given by

$$\begin{bmatrix} M_{11} & M_{12} \\ M_{21} & M_{22} \end{bmatrix} = D_0^{-1} \left[\prod_{j=1}^{N} D_j P_j D_j^{-1} \right] D_R$$

$$D_j = \begin{bmatrix} 1 & 1 \\ 1/Z_j & -1/Z_j \end{bmatrix} \tag{4}$$

$$P_j = \begin{bmatrix} e^{-ik_j d_j} & 0 \\ 0 & e^{ik_j d_j} \end{bmatrix}$$

D_j is called the dynamical matrix and arises from the continuity conditions on the pressure and the displacement normal to the interface between media $j - 1$ and j. P_j is the kinematical matrix inside the layer j. The transmittance T is given by the ratio of the Poynting power flow of the transmitted wave to that of the incident wave, and is given in terms of the transfer matrix, by

$$T = \frac{Z_0}{Z_R} \left| \frac{1}{M_{11}} \right|^2 \tag{5}$$

The 1s-counting sequence is generated by counting the number of 1s in the binary representation of the natural numbers. For example, the natural numbers 0, 1, 2, 3, 4, 5, 6, 7, etc can be represented, respectively, in the binary numeral system by 0, 1, 10, 11, 100, 101, 110, 111, etc. Counting the number of 1s in each binary representation, we are left with the sequence P = 0, 1, 1, 2, 1, 2, 2, 3, ... This sequence can be constructed recursively. For step zero, we take 0 as the initiator. For step one, we sum 1 to 0 and we make a concatenation with the initiator, obtaining 0, 1. For step two, we again sum

1 to these last numbers and we make again a concatenation with the preceding numbers to obtain 0, 1, 1, 2. For step three, we sum 1 once more to the four last numbers and we make again the concatenation to obtain 0, 1, 1, 2, 1, 2, 2, 3, and so on. Every step doubles the length of the sequence. For step S, the number of terms of the sequence is 2S. Since the sequence can be constructed recursively, it has the property of self-similarity. For example, if we take the sequence by pairs (0, 1), (1, 2), (1, 2), (2, 3), . . ., this sequence also follows the original sequence, and also for quartets, octets, etc. Also, if we underline every second term of the sequence, we can reproduce the original sequence. The structure we propose is made of N = 2S layers, each one having an acoustic impedance Z given by

$$Z_Q = Z_1 + (\Delta)(Q) \tag{6}$$

where Z_1 is an initial impedance for the first layer, Δ a small increment, S is the number of the step for the sequence and Q takes the values of the terms of the sequence for a given value of S, This structure could be fabricated using the technology of porous silicon where the acoustic properties depend on the porosity.

3. Results

We consider that the structure is situated in water, i. e. media 0 y R are water, with Z_0=1.5 MRayl, we take Z_1=2.2 MRayl, and Δ=0.5. The thickness of the layers is a quarter wavelenght $\lambda_j/4$, being λ_j the wavelength of the sound for each j layer. Fig. 1 presents the transmission spectrum T dependent on the frequency $f=\omega/2\pi$ for a structure of 8 layers; we see regularity for the spectrum although is difficult to see more characteristics due to the small number of layers. Fig. 2 shows the transmittance for N=32 layers; in this case the self-similarity is clear for the spectrum because if one decreases the transmittance minima with magnitude around 0.5, their structure is similar to the minima around 0.3. If we decrease their magnitude, they are in turn similar to the minima close to 0.04.

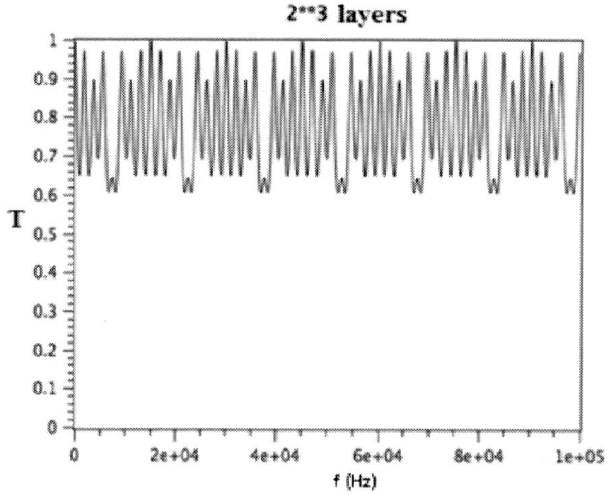

Figure 1 Transmittance for a structure with N = 8 layers.

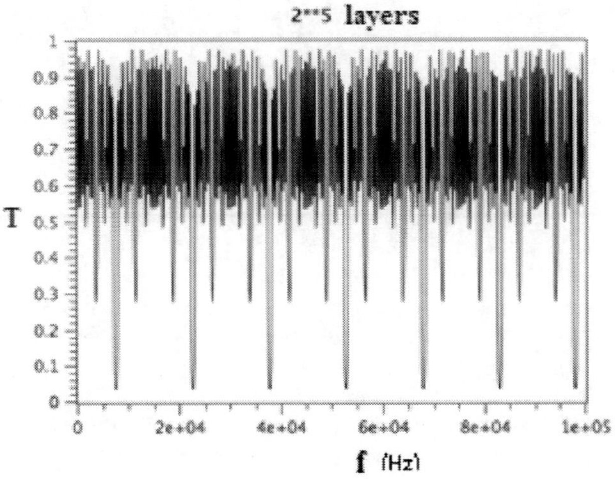

Figure 2 Transmittance for a structure with $N = 64$ layers.

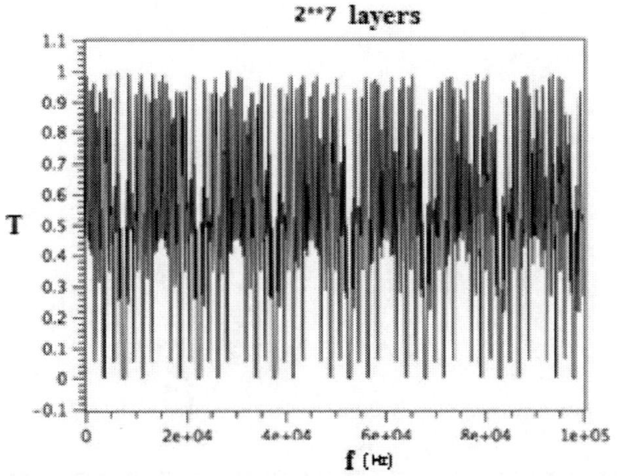

Figure 3 Transmittance for a structure with N = 128 layers.

Fig. 3 presents the transmittance for a structure of 128 layers. This more complex spectrum shows more self-similarity for more scales than that of Fig. 2. It is clear that the spectrum is a prefractal object, if we take a small part of the spectrum, is a reduced copy of the whole spectrum. The perfect fractality is obtained when the number S of steps goes to infinity, i. e., when the number N of layers is infinite.

4. Conclusions

Using the transfer matrix method, we have studied theoretically the transmittance spectrum for a multilayer acoustic structure where the characteristic acoustic impedance follows the values of the self-similar 1s-counting sequence, formed by the quantity of 1s in the binary representation of the natural numbers. We found that the spectrum has also properties of self-similarity.

References

[1] Shechtman D S, Blech I, Gratias D, and Cahn J W 1984 Metallic phase with long–range orientational and no translational symmetry, *Phys. Rev. Lett.* **53** 1951

[2] Merlin R, Bajema K, Clarke R, Juang F Y and Bhattacharya P K 1985 Quasiperiodic GaAs–AlAs heterostructures *Phys. Rev. Lett.* **55** 1768

[3] Velasco V R 2002 Electronic properties of Fibonnaci quasi-periodic heterostructures *phys. stat. sol.* (b) **232** 71

[4] Lavrinenko V, Zhukovsky S V, Sandomirski K S and Gaponenko S V 2002 Propagation of classical waves in nonperiodic media: Scaling properties of an optical Cantor filter *Phys. Rev.* E **65** 036621

[5] Craciun F, Bettucci A, Molinari E, Petri A and Alippi A 1992 Direct experimental observation of fracton mode patterns in one-dimensional Cantor composites *Phys. Rev. Lett.* **68** 1555

[6] Shen M and Cao W 2000 Acoustic band gap formation in a periodic structure with multilayer unit cells *J. Phys. D: Appl. Phys.* **33** 1150

[7] http://oeis.org/A000120 *Consulted* March 15, 2012

[8] Saldaña X I, López-Cruz E and Contreras-Solorio D A 2009 Self-similar optical transmittance for a deterministic aperiodic multilayer structure, *J. Phys.: Condens. Matter* **21** 155403

IOP Conf. Series: Materials Science and Engineering **45** (2013) 012031 doi:10.1088/1757-899X/45/1/012031

Transmission of electrons with flat passbands in finite superlattices

A H Barajas-Aguilar, K A Rodríguez-Magdaleno, J C Martínez-Orozco, A Enciso-Muñoz and D A Contreras-Solorio

Unidad Académica de Física. UAZ. Calzada Solidaridad esq. Paseo la Bufa S/N, C.P. 98060. Zacatecas, Zac. México

E-mail: dacs10@yahoo.com.mx

Abstract. Using the transfer matrix method and the Ben Daniel-Duke equation for variable mass electrons propagation, we calculate the transmittance for symmetric finite superlattices where the width and the height of the potential barriers follow a linear dependence. The width and height of the barriers decreases from the center to the ends of the superlattice. The transmittance presents intervals of stopbands and quite flat passbands.

1. Introduction

The subject of propagation of any kind of waves in layered structures is very important. One of their important applications is as reflector structures and as filters. The search for energy, electronic, optical and acoustic filters is an interesting and active field. Formerly the transmittance for a superlattice where the barriers height is modulated by a Gaussian was calculated and very flat passbands are obtained [1]. In this work we propose a symmetric structure where the width and the height of the potential barriers follow a linear profile. The highest and widest barrier is at the center of the superlattice and the lowest or narrowest ones are at the ends. The idea behind using these types of structures is that it provides a slowly varying potential for the layers, which can improve the transmission of electrons through the multilayer system. For the calculations we use the transfer matrix method and the Ben Daniel-Duke equation for variable mass electrons [2]. The spectrum of transmittance of the structure presents stopbands and nearly flat transmission bands of energy. We compare this transmittance with that produced by a regular superlattice where all the barriers have the same height and width, and also for an inverted structure where the barriers width increases from the center to the ends.

2. Method

The propagation of the electrons in the structure is described by the Ben Daniel-Duke equation, appropriate for electrons with variable effective mass m^*.

$$-\frac{\hbar^2}{2}\frac{d}{dz}\left\{\left(\frac{1}{m^*(z)}\right)\frac{d\psi(z)}{dz}\right\}+V(z)\psi(z)=E\psi(z) \tag{1}$$

The boundary conditions are the continuity of $\psi(z)$ and $1/m^*(d\psi/dz)$. In order to solve the previous equation for our structures, we use the formalism for transfer matrix and transmission coefficient of Pérez-Alvarez and Rodriguez-Coppola [3].

We consider a superlattice made with the materials *GaAs* / $Al_xGa_{1-x}As$. We take the *GaAs* gap as 1.42 eV. We consider an *AlAs* concentration until x=0.45 for which *AlGaAs* has still a direct gap of 1.98 eV. The conduction band offset for the GaAs / *AlGaAs* interface is 0.6. The maximum height which can be taken for the barriers is 0.33 eV corresponding to a concentration x=0.45. The width of barriers and wells is given in monolayers, one monolayer (ML) has a thickness of 2.825 Å. In order to calculate the electron effective mass m for the alloy, we use the virtual crystal approximation [4], with

$$\frac{1}{m_j} = \frac{x_j}{m_A} + \frac{1-x_j}{m_G} \tag{2}$$

m_A and m_G being the electron effective masses for pure $AlAs$ and $GaAs$, which we take respectively as 0.15 and 0.067 of the free electron mass. With this approximation the concentration is given by $x = V / 0.733$, where V is the height of the barriers for the superlattice.

3. Results

We consider superlattices where the height or the width of the barriers follows a symmetric linear variation where the highest or widest barrier is at the center, and the lowest or narrowest ones are at the ends of the structure. The well width is constant for a particular structure. Fig. 1 presents the transmittance versus energy of the incident electrons for two types of structures with 13 barriers. V_{max}, V_{min}, are respectively the maximum and minimum height of barriers, while h_{max}, h_{min} are the maximum and minimum width of barriers considered. One structure has only variation of the barriers height and the other one has both variations of height and width.

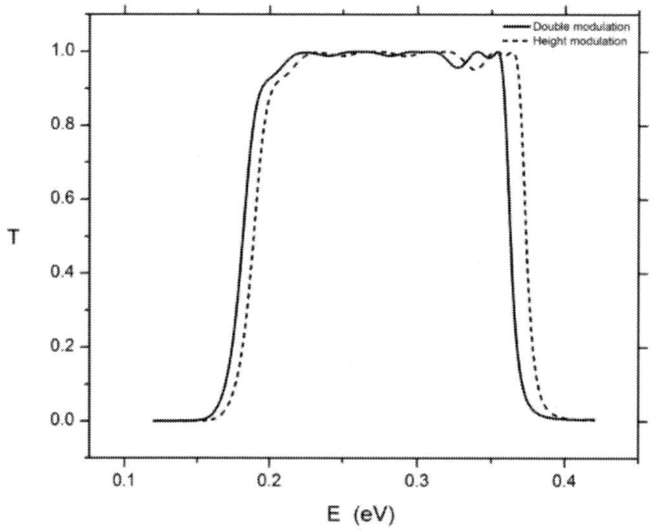

Figure 1. Transmittance vs. energy, 13 barriers. Dashed curve: only height barriers modulation, V_{max}=0.33 eV, V_{min}=0.047 eV. Solid curve: also with width barriers modulation, h_{max}=5 ML, h_{min}=2 ML. Well width=22 ML.

Fig. 2 shows also the transmittance for two types of structures with 13 barriers with another set of parameters for the barriers and wells. One structure has only variation of the barriers width and the other one has both variations of width and height. The flat passbands are outstanding for Figs. 1 and 2.

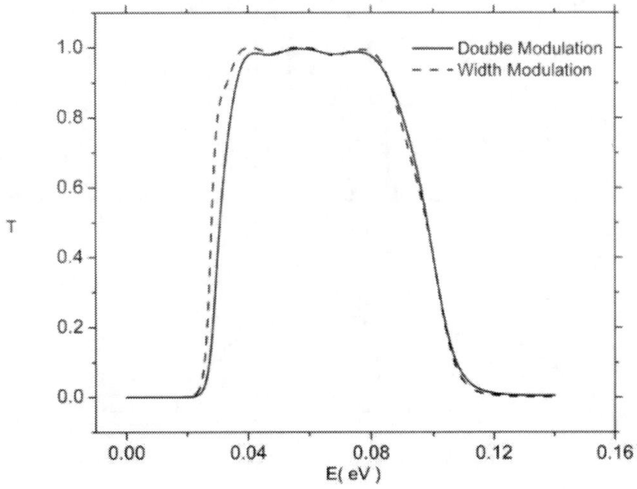

Figure 2. Transmittance vs. energy, 13 barriers. Dashed curve: only width barriers modulation, hmax=7 ML, hmin=1 ML, V=0.15 eV. Solid curve: also with height barriers modulation, Vmax=0.15 eV, Vmin=0.09 eV. Well width=20 ML.

Fig. 3 shows in solid line the transmission for a superlattice with barriers of height of 0.20 eV and variable width; the wells width is 20 ML. For comparison we present in dashed line the transmission of a regular superlattice where the 13 barriers have the same width of 7 ML.

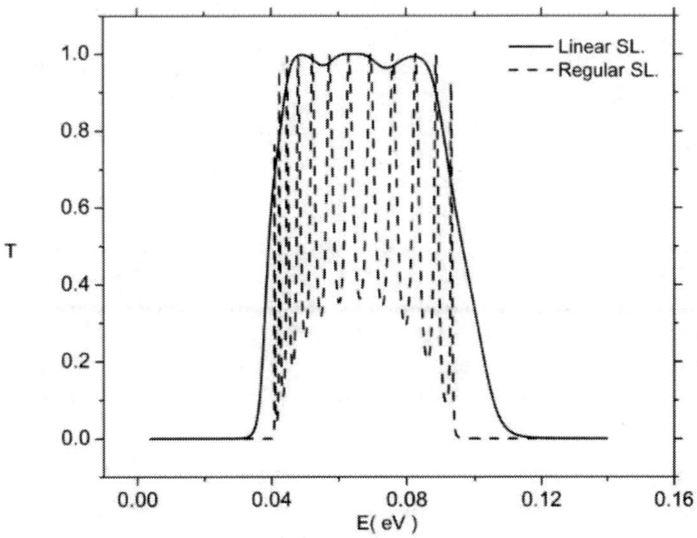

Figure 3. Transmittance vs. energy, 13 barriers, V=0.20 eV, wells width=20 ML. For the solid curve the barriers width is hmax=7 ML, hmin=1 ML. The dashed curve is the transmittance for a regular structure with the same width of 7 ML for all the barriers.

We stress the flatness of the passband for the superlattice with a linear variation of the barriers width in opposition to the transmission for the regular structure. The passband for the regular superlattice has oscillations which are the curves of resonance for the eigenenergies due to the 12 wells.

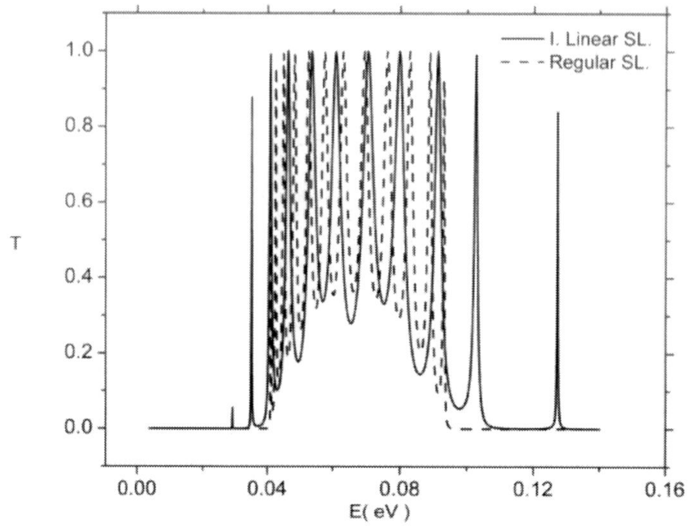

Figure 4. Transmittance vs. energy, 13 barriers, V=0.20 eV, wells width=20 ML. Solid curve is transmittance for the inverted linear superlattice with h_{min}=1 ML at the center of the structure and h_{max}=7 ML at the ends. Dashed curve is transmittance for the regular superlattice of Fig. 3.

The decrease of the barrier width from the center toward the sides of the structure causes a drop of the lifetime Δt for the eigenstates, due to the fact that the electrons tunnel more easily. For the uncertainty principle $\Delta E\ \Delta t \geq \hbar/2$, the resonance bandwidth ΔE increases and the passband, which is the envelope of the resonance curves, becomes flatter. Finally, Fig. 4 presents the transmittance for the regular structure of Fig. 3 and the transmittance for an inverted superlattice with an opposite linear variation of the barriers width, where the narrowest barrier is at the center of the structure while the widest ones are at the ends. For this last structure the electrons are more confined that in the regular structure and the resonance peaks are narrower.

4. Conclusions

We have made a study of the electrons transmittance for a finite superlattice where the height and width of the potential barriers follows a linear profile. The highest or the widest barrier is at the center of the structure and the lowest or thinnest ones are at the ends. The transmittance presents flat transmission and reflection bands. This type of structure could have applications as energy filters for electrons, allowing electrons of a selected energy interval to pass through. This superlattice could also be suitable for superlattice solar cells. For this purpose we plan to do calculations for these structures subject to an electric field.

References

[1] Domínguez-Adame F, Diez E and Bellani V (1999) *J. Appl. Phys.* **85**(7) 3916
[2] Ben Daniel D J and Duke C B 1966 *Phys. Rev.* **152**, 683
[3] Pérez-Alvarez R and Rodríguez-Coppola H 1988 *phys. stat. sol.* (b) **145**, 493
[4] Singh J 1993 *Physics of semiconductors and their heterostructures* (McGraw-Hill)

Third Congress on Materials Science and Engineering (CNCIM-Mexico 2012) IOP Publishing
IOP Conf. Series: Materials Science and Engineering **45** (2013) 012032 doi:10.1088/1757-899X/45/1/012032

Optical transmittance of a multilayer structure with Gaussian modulation of the refractive index

J Madrigal-Melchor, A Enciso-Muñoz, and D A Contreras-Solorio

Unidad Académica de Física de la UAZ, Apartado Postal C-580, 98060 Zacatecas, Zac., México

Email: dacs10@yahoo.com.mx

Abstract. Using the transfer matrix method we calculate the electromagnetic transmittance for a multilayer dielectric structure, where the refractive index of the layers with odd numbering follows a Gaussian distribution of values, while the inserted even layers have a constant refractive index. The transmission spectrum presents intervals of stopbands and very flat passbands.

1. Introduction

Layered structures are very important in the subject of propagation of any kind of waves. They can have applications as reflector structures and as filters and are also known as one-dimensional photonic structures. The search for energy, electronic, optical and acoustic filters is an interesting and active field. There are proposals of energy band-pass filters using quantum superlattices with a Gaussian potential profile [1]. These layered structures allow the incident electrons to be nearly totally transmitted when the impinging electron energy is in a passband. Also, a complete reflection occurs when the impinging energy is in a stopband. The characteristics of the bands can be adjusted modifying the parameters of the superlattice and of the Gaussian distribution. Also, layered omnidirectional optical mirrors have been proposed where the refractive index varies according to the envelope of a Gaussian function, using normal materials [2] and using also metamaterial [3]. These systems were proposed as omnidirectional mirrors but their properties as optical filters were not investigated. Likewise, there is a proposal of a layered acoustic filter where the acoustic impedance of the layers is modulated by a Gaussian function [4]. In this case, the frequency dependent acoustic transmission presents stopbands and very flat passbands. For light, the refractive index is one of the main optical properties of a material. Here we study the frequency dependence of the electromagnetic transmittance for a multilayer structure where the refractive index of the layers with odd numbering follows a Gaussian distribution of refractive indices. The inserted even layers have a constant refractive index. A Gaussian profile provides a slowly varying refractive index for the layers, which can improve the transmission of the structure. For the calculations, we follow the transfer matrix method used in [5]. Our results show that the transmittance spectrum has well defined stopbands and nearly flat passbands. This type of structure can be constructed using layers of porous silicon. This material is widely used for the fabrication of optical devices where the refractive index can be varied through a variation of the porosity [6, 7].

IOP Conf. Series: Materials Science and Engineering **45** (2013) 012032 doi:10.1088/1757-899X/45/1/012032

2. Theoretical model and calculation method

We consider a structure of N plane dielectric layers perpendicular to the x axis. Each j layer has a width d_j and refractive index n_j. Each medium j is homogeneous, isotropic and lossless. We have an electromagnetic plane wave in the xz plane, coming from the left in a medium with refraction index n0. The wave is incident normal to the structure. At the right side of the structure, the wave propagates in a medium S with refraction index n_S. The electric field in the incident medium, in the layers, and in the substrate is given by

$$\vec{E}_j(x,t) = \vec{E}_{1j} e^{i(k_j x - \omega t)} + \vec{E}_{2j} e^{i(-k_j x - \omega t)} \quad (1)$$

Here the first and second terms on the right-hand side of Eq. 2 represent waves propagating to the right and to the left in the x direction, k_j is the wave vector, with $j = 0, 1, \ldots, N, S$. t is the time, ω the angular frequency and i is the imaginary unit. Considering only propagation to the right in the medium S, then $E_{2S} = 0$. The corresponding magnetic field can be obtained using the Faraday's Law. Propagation from medium 0 to medium S through the structure can be described by

$$\begin{bmatrix} E_{10} \\ E_{20} \end{bmatrix} = \begin{bmatrix} M_{11} & M_{12} \\ M_{21} & M_{22} \end{bmatrix} \begin{bmatrix} E_{1S} \\ 0 \end{bmatrix}. \quad (2)$$

Where the (2 × 2) transfer matrix is given by

$$\begin{bmatrix} M_{11} & M_{12} \\ M_{21} & M_{22} \end{bmatrix} = D_0^{-1} \left[\prod_{j=1}^{N} D_j P_j D_j^{-1} \right] D_S. \quad (3)$$

$$D_j = \begin{bmatrix} 1 & 1 \\ n_j & -n_j \end{bmatrix}, \quad P_j = \begin{bmatrix} e^{-ik_j d_j} & 0 \\ 0 & e^{ik_j d_j} \end{bmatrix}$$

D_j is the dynamical matrix, which arises from the continuity conditions on the electric and magnetic fields at the interface between media $j-1$ and j. P_j is the propagation matrix inside the j layer. The transmittance T is given by the ratio of the Poynting power flow of the transmitted wave to that of the incident wave, and is given in terms of the transfer matrix by

$$T = \frac{n_s}{n_0} \left| \frac{1}{M_{11}} \right|^2 \quad (4)$$

The refractive index for the layers with odd j is modulated by the Gaussian function

$$n(x) = (n_{\max} - n_{\min}) \exp(-(x - x_0)^2 / \sigma^2) + n_{\min} \quad (5)$$

Where $n_{\max}$ and $n_{\min}$ are the maximum and minimum values considered for the refractive index of the layers. σ is the standard deviation of the Gaussian distribution.

3. Results and discussion

The structure is in vacuum, then $n_0 = n_s = 1$. We take for our calculations a structure with a total number of 45 layers, except for Fig. 3 where the number of layers is varied. The 23 layers with odd numbering j have refractive index modulated by the Gaussian function. We call these ones the Gaussian or odd layers. The 22 layers with even numbering inserted in between the Gaussian layers have the same refractive index as the incidence medium, $n = 1$. The total thickness of the structure is fixed, with a value $\Delta = 1$ in arbitrary units. For the Gaussian function we use the value $\sigma = \Delta/4$, which gives a more efficient transmission for the band-pass. We assume that the structure can be made using porous silicon which has refractive index of 3.5 for long wavelengths [8]. On the other hand, structures with porous silicon reaching $n = 1.2$ can be constructed [8]. We suppose then that $n_{\max} = 3.5$ and $n_{\min} = 1.2$ in Eq. 5. The Gaussian layers have width d_g and the inserted layers have width d_l. The spectra for the multilayer structures consist of passbands of discrete values of eigenfrequencies, separated by stopbands, where there is no transmission of electromagnetic waves. Fig. 1 shows the transmission for three values of the ratio $d_l/d_g = 4, 6, 8$, as a function of $f\Delta/c$, where f is the radiation frequency, and c the speed of light in vacuum.

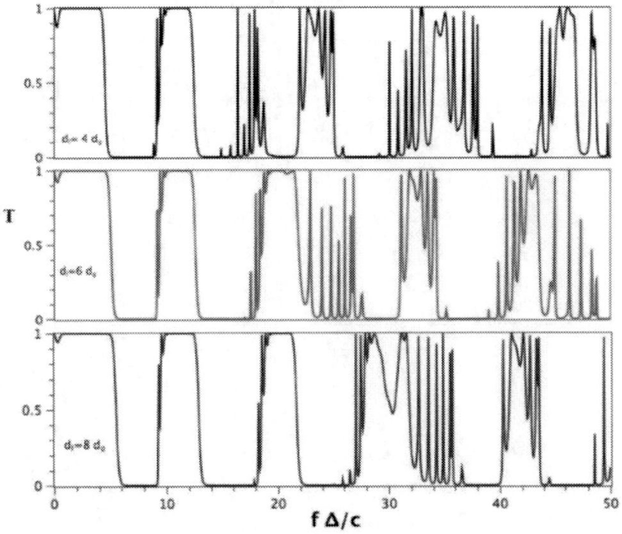

Figure 1 Transmission for a structure with 45 layers, where 23 follow the Gaussian distribution, for three values of the ratio $d_l/d_g = 4, 6$, y 8 with $n_{\min} = 1.2$, $n_{\max} = 3.5$

For the three spectra there is a low passband followed by stopbands and passbands. The practical interest lies in the flat passbands with almost total transmission for a frequency interval. When the ratio d_l/d_g increases, there is better transmission in the passband, their width increases and the size of the stopbands decreases. The reason of the improvement of the transmission, is that the fraction filled by layers of lower index, with $n = 1$, is larger than the fraction of the structure filled by Gaussian layers which have bigger values of refractive index. Fig. 2 shows the transmission for a non Gaussian regular structure, where the layers with odd numbering have a constant value of refractive index, $n = 3$. This regular structure presents the expected gaps of transmission or stopbands, but it does not have flat passbands as the structure with Gaussian profile. A structure with regular profile can work as a mirror, but it would be bad as a filter. Fig. 3 presents the transmission for three structures with different number of layers.

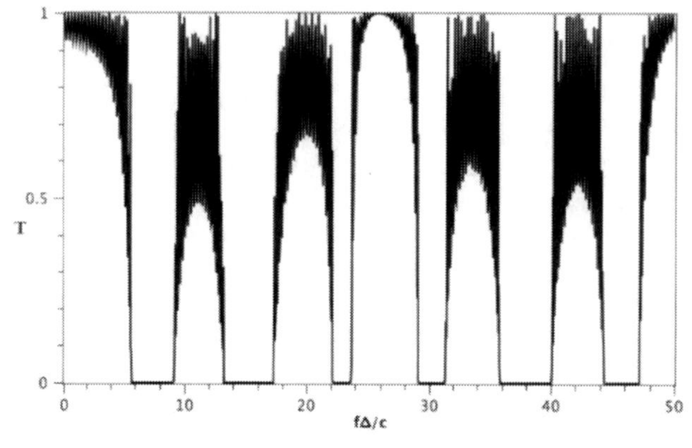

Figure 2 Transmission for a regular structure of 45 layers, where the odd layers have $n = 3.0$, the even ones $n = 1$, and $d_l/d_g = 6$

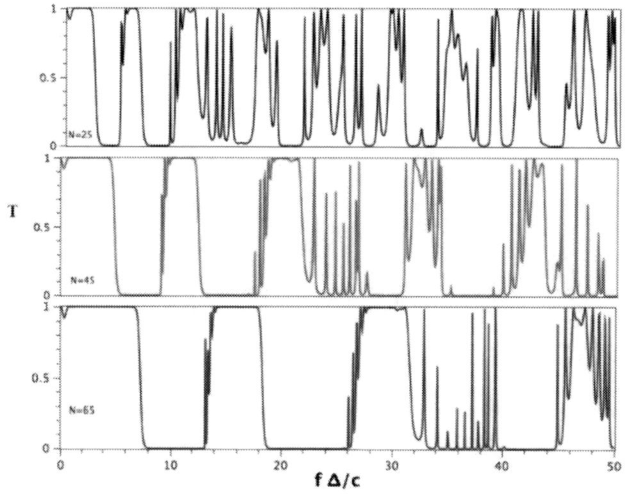

Figure 3 Transmission for structures with 25, 45 and 65 layers using the ratio $d_l/d_g = 6$ and $n_{min} = 1.2$, $n_{max} = 3.5$

When there are more layers in the structure, there is a shift of the bands to higher frequencies and the passbands are wider because the number of eigenfrequencies increases. At the same time, the reflectance for the stopbands is improved due to the fact that there are more layers with large refractive index.

4. Conclusions

Using the transfer matrix method, we studied theoretically a multilayer structure situated in vacuum where the refractive index of the layers follows a Gaussian distribution. These layers are separated also by layers of vacuum. We obtained very flat passbands separated by stopbands. This type of structure can have potential applications as a filter, which allows selected intervals of frequencies to pass through.

References

[1] Domínguez-Adame F, Diez E and Bellani V (1999) *J. Appl. Phys.* **85**(7) 3916

[2] Arriaga J and Saldaña X (2006) Band structure and reflectivity of omnidirectional Si-based mirrors with a Gaussian profile refractive index, J. Appl. Phys. 100(4) 044911

[3] Saldaña Saldaña X, Contreras Solorio D A, Enciso Muñoz A, Madrigal Melchor J (2008) Reflectance of a layered system with a gaussian distribution of the refractive index and inserted metamaterial, *Revista Investigación Científica* **4**(2) 1

[4] Madrigal-Melchor J, Reyes Villagrana R A, Saldaña X, Jelev-Vlaev S and Contreras-Solorio D A (2008) Acoustic layered filter with a Gaussian distribution of impedances, *SLAFES* Book of Abstracts Iguazú, Argentina

[5] P. Yeh 1988 *Optical waves in layered media*, (Ed. Wiley, New York)

[6] Mattei G, Marucci A, Yakovlev V A, and Pagannone M (1998) Porous silicon optical filters for application to laser technology, *Laser Phys.* **8**(3) 755-758

[7] Lorenzo E, Oton C J, Capuj N E, Ghulinyan M, Navarro-Urrios D, Gaburro Z, and Pavesi L (2005) Porous silicon-based rugate filters, *Appl. Opt.* **44**(26) 5415

[8] *Optical Properties of Silicon*, Virginia Semiconductor, Inc. www.virginiasemi.com

AUTHOR INDEX

AUTHOR INDEX